APPLICATIONS OF THE VOLTAMMETRY

Edited by **Margarita Stoytcheva**
and **Roumen Zlatev**

Applications of the Voltammetry
http://dx.doi.org/10.5772/65154
Edited by Margarita Stoytcheva and Roumen Zlatev

Contributors

Ali R. Jalalvand, Maria Valnice Zanoni, Felipe Fantinato Hudari, Michelle Fernanda Brugnera, Francisco J. Heredia, Dolores Hernanz-Vila, M. José Jara-Palacios, M. Luisa Escudero-Gilete, Krzysztof Suchocki, Pipat Chooto, Jose Luis Hernandez-Lopez, Flavio Dolores Martinez-Mancera

Published by InTech
Janeza Trdine 9, 51000 Rijeka, Croatia

Publishing Process Manager Mirena Calmic
Technical Editor SPi Global
Cover InTech Design team

First published May, 2017
Printed in Croatia
Legal deposit, Croatia: National and University Library in Zagreb

Additional hard copies can be obtained from orders@intechopen.com

Applications of the Voltammetry, Edited by Margarita Stoytcheva and Roumen Zlatev
p. cm.
Print ISBN 978-953-51-3215-8
Online ISBN 978-953-51-3216-5

For EU product safety concerns:
IN TECH d.o.o., Prolaz Marije Krucifikse Kozulić 3, 51000 Rijeka, Croatia,
info@intechopen.com or visit our website at intechopen.com.

Meet the editors

Prof. Margarita Stoytcheva has graduated from the University of Chemical Technologies and Metallurgy of Sofia, Bulgaria, with titles Chemical Engineer and Master of Electrochemical Technologies. She has PhD and DSc degrees in chemistry and technical sciences. She has participated in research and teaching in several universities in Bulgaria, Algeria, and France. From 2006 to the present, she has participated in activities of scientific research and technological development and is teaching at the Autonomous University of Baja California at Mexicali, Mexico, as a full-time researcher. Since 2008, she has been a member of the National System of Researches of Mexico, and since 2011, she has been a regular member of the Mexican Academy of Sciences. Her interests and area of research are analytical chemistry and biotechnology.

Dr. Roumen Zlatev is a full-time researcher in the Engineering Institute of the Autonomous University of Baja California (UABC) at Mexicali, Mexico. He obtained his Bachelor's and Master's degrees from the University of Chemical Technology and Metallurgy of Sofia, Bulgaria, and his PhD degree from the National Polytechnic University of Grenoble, France. He was a full-time researcher in the Bulgarian Academy of Sciences and a part-time professor at the Sofia University. In 2005, he accepted the position of full-time senior researcher in UABC.

Dr. Zlatev is a member of the Mexican National System of Researchers and a regular member of the Mexican Academy of Sciences. He has participated in research projects in France, Germany, and Mexico. He is the author of more than 170 publications, book chapters, and reports in scientific congresses and he holds 14 patents in the field of the electrochemical and spectroscopic methods of analysis, corrosion and materials, electrochemical and analytical instrumentation.

Contents

Preface

Voltammetry is an electrochemical technique based upon the measurement of the current at an electrode as a function of the applied time-dependent potential. The wide-ranging applications of voltammetry include studies associated with the characterization of the electrochemical properties of the chemical substances, as well as their quantitative determination.

The Nobel Prize–winning chemist J. Heyrovsky developed the first voltammetric technique, the DC polarography. J.C. Barker made the next significant step with the development of square wave, differential and normal pulse, as well as radiofrequency polarography. Polarography is referred to the voltammetry applied with mercury electrode. The recent achievements in the material science offered materials and technologies allowing the mercury electrode replacement by solid electrodes. This type of electrodes, combined with nanoscale-sized modifiers, converted the voltammetry into a powerful but a low-cost method for the rapid, direct, specific and high-sensitive quantification of a variety of inorganic and organic species.

The present book Applications of Voltammetry is a collection of six chapters, organized in two sections.

The first book section 'Metrology and Chemometrics' (Chapters 1 and 2) is dedicated to the application of mathematical methods, such as multivariate calibration coupled with voltammetric data and numeric simulation to solve quantitative electroanalytical problems. The suggested approaches could be useful for the improvement of the sensitivity and the selectivity of the voltammetric techniques.

The second book section, including Chapters 3–6, is devoted to the electron transfer studies and electroanalytical applications of voltammetry. Chapter 3 comments on the interfacial electron transfer kinetics of the haem group in human haemoglobin molecules and physisorbed on glass-/tin-doped indium oxide substrates. Chapter 4 offers an exhaustive overview on the application of voltammetry to the analysis of dyes, using a variety of techniques and electrodes. Chapter 5 discusses about voltammetric methods suitable for characterizing the antioxidant properties of wine and wine products. Chapter 6 addresses the large spectrum of modified electrodes, applied for determining trace metal ions with improved sensitivity.

The book offers a professional look on the recent trends and advances in voltammetry. All the contributing authors are gratefully acknowledged for their time and efforts.

Margarita Stoytcheva and **Roumen Zlatev**
Universidad Autónoma de Baja California
Instituto de Ingeniería
Mexicali, Baja California, México

Section 1

Metrology and Chemometrics

Chapter 1

Applications of Chemometrics-Assisted Voltammetric Analysis

Ali R. Jalalvand

Additional information is available at the end of the chapter

http://dx.doi.org/10.5772/67310

Abstract

Electroanalytical techniques consist of the interplay between electricity and chemistry, namely the measurement of electrical quantities, such as charge, current or potential and their relationship to chemical parameters. Electrical measurements for analytical purposes have found a lot of applications including industry quality control, environmental monitoring and biomedical analysis. Chemometrics is the chemical discipline that uses mathematical and statistical methods to design or select optimal procedures and experiments and to provide maximum chemical information by analysing chemical data. The use of chemometrics in electroanalytical chemistry is not as popular as in spectroscopy, although recently, applications of these methods for mathematical resolution of overlapping signals, calibration and model identification have been increasing. The electroanalytical methods will be improved with the application of chemometrics for simultaneous quantitative prediction of analytes or qualitative resolution of complex overlapping responses. This chapter focuses on applications of first-, second- and third-order multivariate calibration coupled with voltammetric data for quantitative purposes and has been written from both electrochemical and chemometrical points of view with the aim of providing useful information for the electrochemists to promote the use of chemometrics in electroanalytical chemistry.

Keywords: chemometrics, electroanalytical chemistry, voltammetry, quantification, multivariate calibration

1. Introduction

1.1. Chemometrics

Chemometrics uses statistics, mathematics and formal logic: (i) to provide maximum relevant chemical information by analysing chemical data, (ii) to design or to select optimal experimental procedures and (iii) to obtain knowledge about chemical systems [1]. Chemometric analyses have received good acceptance over the past 20 years due to the study of complex samples by improving existing analytical methods. There is an impressive research related to the development and testing of multivariate algorithms applied to difficult chemical scenarios [2, 3]. The main reason is that the second- and higher-order data are able to deal with unwanted interferences in contrast to zero- and first-order calibrations [4]. Modelling of the unwanted interferences that are not included in the calibration set allows us to accurate determination of the calibrated analytes even in the presence of uncalibrated interferences. The property is a 'second-order advantage' [5], which has a great potential in multicomponent analysis. Second- and third-order multivariate calibrations are gaining widespread acceptance by the analytical community due to their variety of second- and third-order instrumental data that are being produced by modern instruments and due to their appeal from the analytical standpoint.

1.2. Electroanalytical methods

Electroanalytical methods can monitor an analyte by measuring the potential or current in an electrochemical cell containing the analyte [6–9]. It is well known that electrochemical analysis has been benefited from the electronics revolution in two ways: (i) the development of neater, faster and simpler and arguably, competitively affordable instrumentation and (ii) the applicability for rapid analysis. In addition, electrochemistry has a wide range of analytical methods, e.g. polarography, potentiometry, coulometry and voltammetry, which can provide a wide concentration range (from ppb to mg L^{-1} levels).

1.3. Chemometrics in electroanalytical chemistry

The use of chemometrics in electroanalytical chemistry is still in its infancy, and for many years, the application of chemometrics to electroanalytical data had been quite scarce as compared to the case of spectroscopic techniques. The limited use of chemometrics in electroanalytical chemistry is related to the relationship between mathematics and electroanalytical chemistry. In this case, the fundamental corpus are: (i) a hypothetical physicochemical picture of the processes, the transport phenomena and the nature of the measurements; (ii) the numerical solution of the mathematical formulation and (iii) the interpretation of the electroanalytical data and determination of concentrations and constants or whatever [10, 11]. This approach is usually called as hard-modelling, a common approach in electrochemical investigations and by the electrochemists, and is regarded as the real approach. Postulation of a theoretical physicochemical model is difficult because the transport phenomenon, the electrode process and perturbation by an excitation signal are complex. In these cases, the other types of approaches that can provide quite more information about the systems are required.

This alternative or complementary approach can be provided by chemometrics, and it is based on extracted results from statistical analysis of the data. This approach is well known as soft-modelling approach.

Chemometrics has different applications in electrochemical analyses, such as experimental design and optimization, data treatments, sample classification, calibration for determination of concentrations and model identification.

In general, over the past two decades, instrumentation has been significantly developed. This is especially the case for electrochemical instruments. Such progress has provided increasing opportunities for the application of chemometrics in electrochemical analyses. Among the most relevant multivariate methods, multi-way algorithms play an important role in numerous analytical fields [12, 13]. Developing an electrochemical technique by chemometric methods may supply a valuable resource for accurate analyte quantification when the absolute separation is not accomplished, or unexpected components are present in the sample being analysed. Chemometrics will be useful when it is coupled to multi-way calibration, for instance, applying excitation-emission [14], high-performance liquid chromatography with diode array detection (HPLC-DAD) [15–17], flow injection analysis- diode array detection (FIA-DAD) [18, 19], liquid chromatography-attenuated total internal reflectance-Fourier transform infrared spectrometry (LC-ATR-FTIR) [20], liquid chromatography-diode array detection-mass spectrometry (LC-DAD-MS) [20], pH-DAD [21, 22], DAD-kinetics [23] and differential pulse voltammetry [24].

1.4. Required information

1.4.1. Calibration

According to the international union of pure and applied chemistry (IUPAC), calibration is, in a general sense, 'an operation that relates an output quantity to an input quantity for a measuring system under given conditions' [25, 26]. The input quantities of our primary interest, i.e. in analytical calibration, are the concentrations of a sample constituent of interest (the analyte), while the output quantities are analytical signals or responses delivered by analytical instruments (a spectrometer, chromatograph, voltammetric equipment, etc.). Therefore, in this chapter, calibration means the operation of relating instrumental signals to analyte concentrations.

1.4.2. Univariate calibration

A specific case of the general calibration process is the one relating the content of a single analyte in a sample to a single value of an instrumental signal and is called 'univariate calibration'. In analytical chemistry, univariate calibration employs a calibration curve as a general method for the determination of the concentration of a constituent in an unknown sample [26].

1.4.3. Multivariate calibration

A more general calibration process involves the relationship between the concentrations of various constituents in a test sample and multiple measured responses, i.e. multivariate instead of univariate [26, 27]. In contrast to univariate calibration, which works with a single instrumental response measured for each experimental sample, multivariate calibration works with many different signals for each sample. Depending on the instrumental setup, the delivered data for a single sample may have different degrees of complexity. The simplest multivariate data are those produced in vector form, i.e. as a series of responses, which can be placed on top of each other to generate a mathematical object known as a column vector. This object is also referred as having a single 'mode' or 'direction'. Multivariate calibration using vectorial data has given rise to a highly fruitful analytical field that today is routine in many industrial laboratories and process control units [28]. Multiple analytes can be determined simultaneously in the presence of others, possibly unknown constituents, provided they have been properly taken into account during the calibration phase [29, 1].

1.4.4. Multi-way calibration

Multi-way calibration is based on many instrumental signals per sample, which can be meaningfully organized into a certain mathematical object with higher modes than a vector, for example, as a data table or matrix [26, 30].

The most important advantage of multi-way calibration is the fact that analytes can be determined in the presence of unexpected constituents in test samples. It is called the 'second-order advantage'.

Multi-way calibration has interesting advantages relative to other calibration methods. One is the increase in sensitivity, because the measurement of redundant data tends to decrease the relative impact of the noise in the signal. Selectivity does also increase, because each new instrumental mode, which is added to the data, contributes positively to the overall selectivity. Still another one is the possibility of obtaining qualitative interpretation of chemical phenomena through the study of multi-way data, in a much better way than with univariate or first-order data.

1.4.5. Nomenclature for data and calibrations

In algebraic jargon, a scalar is a zeroth-order object, a vector is first order, a matrix is second order, etc. A nomenclature exists for the different calibrations, based on the measurement of data of various orders for a single sample: zeroth-order calibration is equivalent to univariate calibration, first-order multivariate calibration is equivalent to calibration with vectorial data per sample, second-order multivariate calibration is equivalent to calibration with matrix data per sample and third-order multivariate calibration is equivalent to calibration with three-dimensional data arrays per sample. The list may continue with data arrays with additional modes per sample [26]. On the other hand, with data for a group of samples it is possible to create an array having an additional mode, the sample mode. For example, univariate measurements for several samples can be grouped to form a vector, first-order data can be placed adjacent to each other to create a matrix, etc. On the basis of above fact, an alternative nomenclature has been developed, in which calibrations are named according to the number of ways (modes) of an array for a sample set. Thus, zeroth-order calibration is also one-way calibration, first order is two-way, second order is three-way, etc., and it is customary to name all calibration methodologies involving second- and higher-order data (i.e. three-way and beyond) as multi-way calibration, which is thus a subdivision of multivariate calibration. **Figure 1** provides a compact view of the hierarchy of data and calibrations.

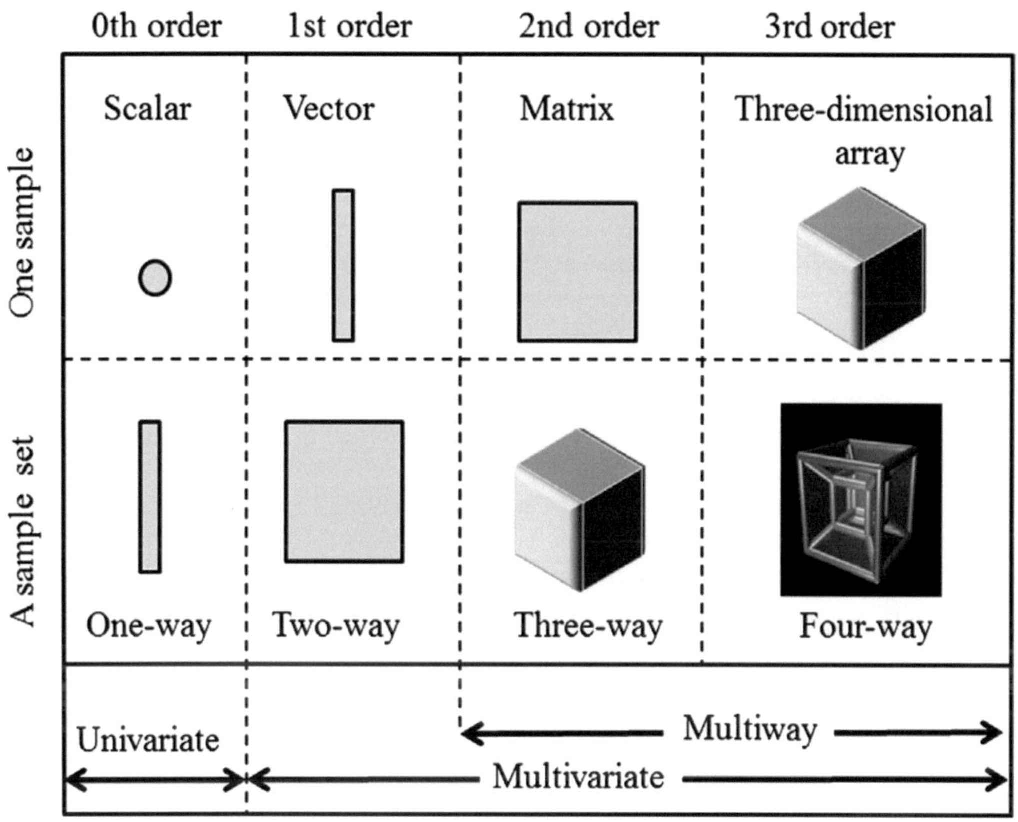

Figure 1. Hierarchy of data illustrating the nomenclatures based on the concept of 'order' and 'ways'. Top: Data for a single sample. Bottom: Data for a set of samples [26].

1.5. Linearity and nonlinearity of second- and third-order data

1.5.1. Second-order data

1.5.1.1. Trilinear data

When a second-order data array is processed, it is vital to meet the so-called trilinearity condition. A three-way data array can be modelled by the following expression:

$$X_{ijk} = \sum_{i=1}^{N} a_{in} b_{jn} c_{kn} + E_{ijk} \tag{1}$$

where N is the total number of chemical constituents generating the measured signal, a_{in} is the relative concentration or score of component n in the ith sample, and b_{jn} and c_{kn} are the intensities in the instrumental channels (or data dimensions) j and k, respectively. The values of E_{ijk} are the elements of the three-dimensional array **E**, representing the residual error, and having the same dimensions as **X**. The column vector a_n is collected in the scores matrix **A**, whereas vectors b_n and c_n are collected in the loading matrices **B** and **C** (usually b_n and c_n are normalized to unit length). For a three-dimensional data array, the signal must be linearly related to concentration and the component profiles must be constant across the different samples [31].

1.5.1.2. Non-trilinear data

To evaluate the linearity of a three-way data array should first consider its basic ingredients, i.e. the individual data matrices and whether they are bilinear or not. In case they are bilinear, a further subdivision can be made on the existence and number of trilinearity-breaking modes: (i) when one of the data modes is non-reproducible and breaks the trilinearity, the data are not trilinear, but can be unfolded into a bilinear augmented matrix and (ii) when both data modes are trilinearity breaking, the data are not trilinear and cannot be unfolded into a bilinear augmented matrix. To distinguish these two latter non-trilinear data types, we propose to call them non-trilinear Type 1 and non-trilinear Type 2, respectively. Finally, in case the individual matrices are non-bilinear, we have a fourth data type that we may call non-trilinear Type 3. There is no point in further dividing Type 3 data according to the number of non-reproducible modes, since the former are neither trilinear nor unfoldable to an augmented bilinear matrix [26, 31]. **Figure 2** illustrates the classification of three-way data for a sample set.

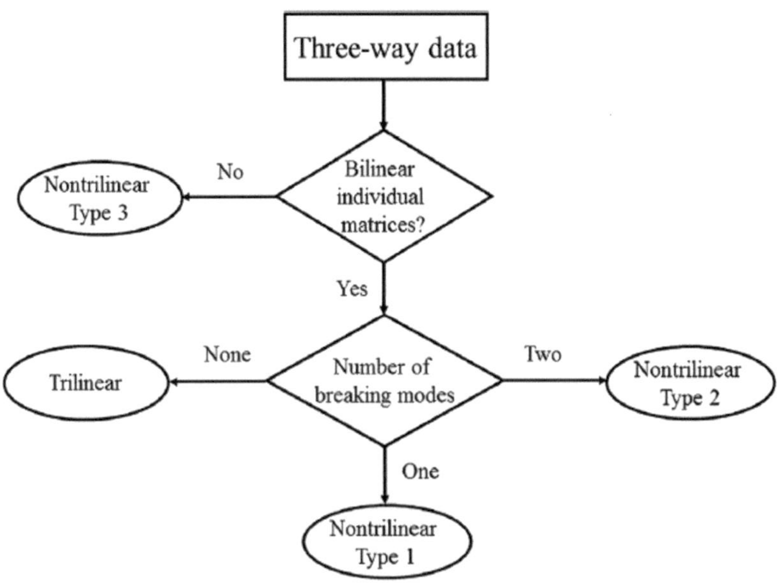

Figure 2. Classification tree for three-way data for a set of samples, according to whether the individual data matrices are bilinear or not and to the number of trilinearity-breaking modes [26].

1.5.2. Third-order data

1.5.2.1. Quadrilinear data

Quadrilinearity of a four-way array can be defined by the extension of Eq. (1), including an additional mode: the sample mode. A four-way data array obtained by 'joining' three-dimensional data arrays for a sample set is a quadrilinear if its elements can be thought to be obtained through:

$$X_{ijkl} = \sum_{n=1}^{N} a_{in} b_{jn} c_{kn} d_{\mathrm{ln}} + E_{ijkl} \tag{2}$$

where all symbols are as in Eq. (1), with d_{in} describing the changes in constituent concentrations along the sample mode. A requirement for quadrilinearity of a data array for a sample set is that the three instrumental profiles for each constituent are equal for all samples [26].

1.5.2.2. Non-quadrilinear data

Quadrilinearity may be lost if one or more modes behave as quadrilinearity-breaking mode, in the sense that constituent profiles change from sample to sample along this mode. In the present case, there might be one, two, or three quadrilinearity-breaking modes. Hence, a pertinent classification of non-quadrilinear third-order/four-way data would be in types 1, 2 and 3, respectively. On the other hand, intrinsically non-trilinear data for each sample for reasons of mutual correlations among the phenomena in the different data modes will be classified as non-quadrilinear of type 4 [26]. **Figure 3** illustrates a classification tree.

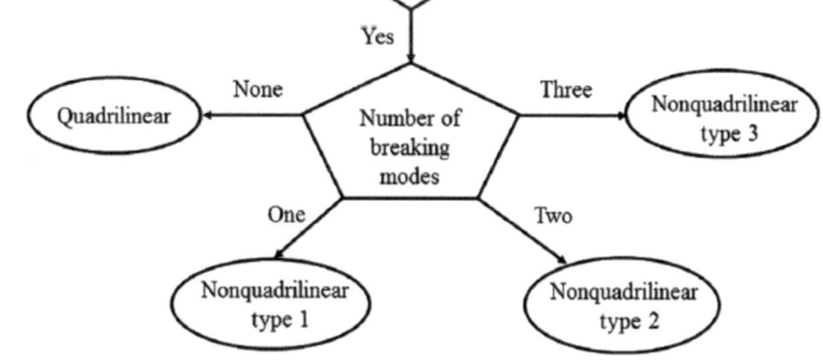

Figure 3. Classification tree for four-way data for a set of samples, according to whether the individual three dimensional arrays data are trilinear or not and to the number of quadrilinearity-breaking modes [26].

1.6. Algorithms

1.6.1. First-order algorithms

The standard models are principal component regression (PCR) and partial least-squares (PLS) analyses, although a number of different algorithms such as continuum power regression (CPR), multiple linear regression-successive projections algorithm (MLR-SPA), robust continuum regression (RCR), partial robust M-regression (PRM), polynomial-PLS (PLY-PLS), spline-PLS (SPL-PLS) and radial basis function-PLS (RBF-PLS) exist. However, when the data behave in a non-linear manner with respect to the analyte, a different approach is needed, such as an artificial neural network (ANN) or a least-squares support vector machine (LS-SVM).

1.6.2. Second-order algorithms

Suitable algorithms for analysing second-order data are parallel factor analysis (PARAFAC) [32], the generalized rank annihilation method (GRAM) [33], direct trilinear decomposition (DTLD) [34], multivariate curve resolution-alternating least squares (MCR-ALS) [35], bilinear least squares (BLLS) [36, 37] and alternating trilinear decomposition (ATLD) [38] and its variants (self-weighted alternating trilinear decomposition (SWATLD) [39] and alternating penalty trilinear decomposition APTLD [40, 41]).

Rearranging the second-order data to vectors and applying a first-order algorithm such as unfolded-principal component regression (U-PCR) and unfolded-partial least squares (U-PLS) [42] is an alternative to working with second-order data. Another alternative which is a genuine multi-way method is multi-way partial least squares (N-PLS) [43]. These three methods can obtain the second-order advantage by coupling of them to residual bilinearization (RBL) [44, 45].

1.6.3. Third-order algorithms

Suitable quadrilinear models for third-order data are including PARAFAC, trilinear least-squares (TLLS) with residual trilinearization (RTL) [46] and alternating penalty quadrilinear decomposition (APQLD) [47]. However, models allowing for deviations of multilinearity in one way or another are including PARAFAC2 (a variant of PARAFAC that allows profile variations in one of the data dimensions from sample to sample) [48], PARALIND (PARAFAC for linearly dependent systems) [49], MCR-ALS [50], non-bilinear rank annihilation (NBRA) [51], bilinear least squares (BLLS) extended to linearly dependent systems [22], U-PLS [42], N-PLS [52], non-linear kernel-PLS [53] and artificial neural networks (ANN) [54, 55]. To achieve the second-order advantage, BLLS, PLS and ANN should be combined with RTL [44–46, 55–59].

1.7. Generation of second- and third-order electrochemical data

Differential pulse voltammetry (DPV) is the most frequently used technique for generation of second- and third-order electrochemical data. The second- or third-order data could be obtained via changing one or two of the instrumental parameters of DPV [60]. The theory behind the proposed procedure will be briefly discussed. The current signal intensity in DPV can be obtained using the following equations [61]:

$$O + ne \rightarrow \text{Red} \tag{3}$$

$$\delta_i = \frac{nFA\, D_O^{1/2}\, C_O^*}{\pi^{1/2}(\tau - \tau')^{1/2}} \left[\frac{P_A(1-\sigma^2)}{(\sigma + P_A)(1 + P_A \sigma)} \right] \tag{4}$$

$$P_A = \xi \exp\left[\frac{nF}{RT}\left(E + \frac{\Delta E}{2} - E^{0'} \right) \right] \tag{5}$$

$$\sigma = \exp\left(\frac{nF}{RT} \frac{\Delta E}{2} \right) \tag{6}$$

$$\xi = \left(\frac{D_O}{D_{Red}} \right)^{1/2} \tag{7}$$

where O and Red are species involved in the electrode reaction (Eq. (1)), n is the number of electrons involved in the electrode reaction, F is the Faraday's constant, A is the electrode area, D_O and D_{Red} are the diffusion coefficients of O and Red species, respectively, C_O^* is the concentration of O species at the electrode surface, R is the gas constant, T is the temperature, ΔE, E and $E^{0'}$ are the pulse height, potential and formal potential of the electrode, respectively, τ and τ' are, the pulse duration or pulse time and starting time of potential pulse, respectively. For an electrochemical reaction, a data vector can be produced by sweeping the potential at constant ΔE and τ. Applying a different ΔE and sweeping potential at the constant τ, produces different data vectors. By the same way, third-order voltammetric data could be obtained by sweeping potentials at different pulse durations and pulse heights [62]. Literature survey shows that changing ΔE can cause non-linearity in DPV data while changing τ does not cause non-linearity [60].

1.8. Data pre-processing

1.8.1. Shift correction or data alignment

Linearity is a property assumed by multivariate linear calibration algorithms. However, in many electroanalytical situations, slight deviations from linearity could be observed such as in the presence of interactions among components. Generally, non-linear signals can cause signal shifts, peak broadening or increase of the peak height. Any of these problems hinders the application of multi-linear data processing algorithms. Such problems become more complicated when signals are overlapping. Therefore, aligning the voltammograms is an important step that should be performed before the application of multi-linear algorithms. The data alignment is based on digitally moving a voltammogram towards a reference voltammogram, with certain objective function such as correlation coefficient, residual fit, similarity index, etc. which indicate the quality of the matching process. The most algorithms used for data alignment require a reference voltammogram, to which all the remaining ones are aligned. Suboptimal choice of the template could affect the alignment results [63]. The most frequently used alignment algorithms are correlation optimized warping (COW) [64], interval correlation optimised shifting (icoshift) [65], Gaussian peak adjustment (GPA) [66], Gaussian peak adjustment with transversal constraints (GPA2D) [67], asymmetric logistic peak adjustment (ALPA) [68], *shiftfit* [69, 70] and *pHfit* [71].

1.8.1.1. COW

For understanding a detailed description about the mathematical aspects of COW algorithm the reader is referred to Ref. [64].

1.8.1.2. *icoshift*

For understanding a detailed description about the mathematical aspects of icoshift algorithm the reader is referred to Ref. [65].

1.8.1.3. *shiftfit, pHfit, GPA, GPA2D and ALPA*

The *shiftfit* corrects the data matrix from the signal movements and for this purpose, it optimises by least squares of the potential shift of every pure voltammogram with respect to a reference position. To correct the potential shift, there are *peakmaker, shiftcalc* and *shiftfit* functions. *Peakmaker* generates a Gaussian peak as an initial estimation of the pure voltammograms. The *shiftcalc* function displaces every signal in every experimental voltammograms matrix for a given potential shift ΔE. The *shiftfit* function iteratively optimises the values of ΔE to generate a matrix (I_{cor}) in which all signals remain at the fixed potentials stated in the pure voltammograms matrix (V_0) [69, 70].

The *pHfit* algorithm can solve more intricate systems like those encountered in voltammetric pH titrations by imposing a shape restriction to the movements of the signals as a function of potential, by means of adjustable sigmoid or linear functions [71].

After *shiftfit* and *pHfit,* the GPA algorithm based on a new strategy, parametric signal fitting (PSF), was proposed for the chemometric analysis of voltammetric data when the pure signals do not maintain a constant shape [66]. This is based on the fitting of parametric functions to reproduce the shape of the signals. As a first approach, two Gaussian functions are fitted, one at each side of the signal, and the parameters are least-squares optimised. Such parameters determine not only the height and position of the signals (as in the algorithms above) but also the width at both sides of the maximum. It is important to note that, unlike *shiftfit* and *pHfit,* the use of Gaussian functions restrict the GPA exclusively to peak-shaped signals. Moreover, it must be remarked that, despite fitting of Gaussian peaks which has been already used in some situations such as the resolution of UV-vis spectra, in such approaches the symmetric character of the Gaussian function prevents an appropriate treatment of asymmetric signals. In the proposed method, the use of two separated Gaussian functions at both sides of the maximum (sharing the same height and position but different widths) is a new and simple solution for the fitting of asymmetric peaks. **Figure 4** summarizes the main steps of the fitting procedure.

A new method, GPA2D, was developed as a significant improvement of the GPA which includes, for the first time, transversal constraints to increase the consistency of the resolution along the different signals of a voltammetric dataset [67]. The aim of GPA2D is to extract a physicochemical sense to the evolution of the signals and their shifts along the experimental axis. The imposition of the transversal constraints makes this method more powerful for the analysis of voltammetric data, especially if they are non-bilinear. **Figure 5** shows the main structure of the operation program, which is based on a common GPA procedure with two alternative intermediate paths depending on the kind of transversal constraint to be applied (signal shift evolution or equilibrium).

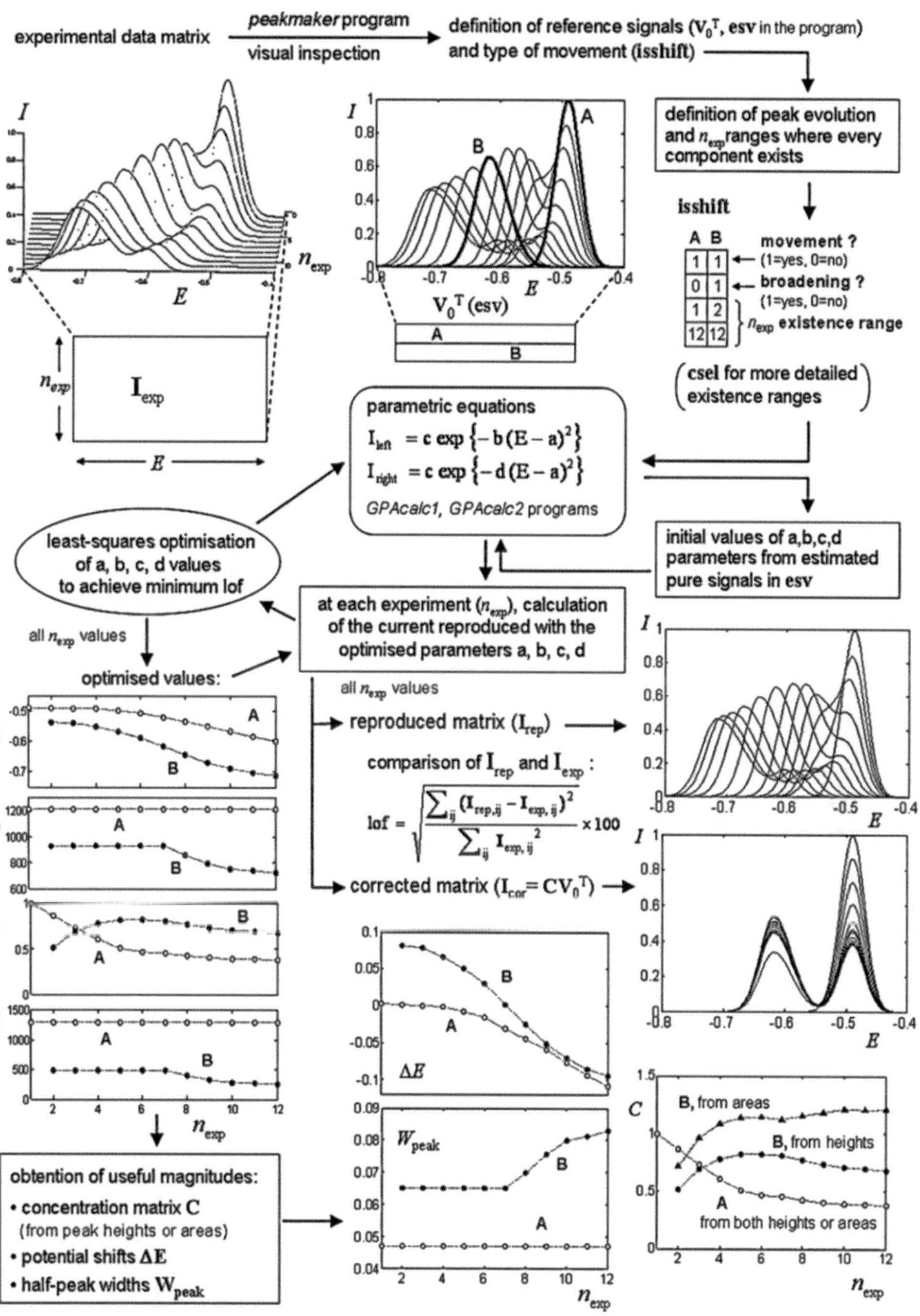

Figure 4. Flowchart of the GPA [67].

The asymmetric logistic peak adjustment (ALPA) was developed as a new function for the PSF of highly asymmetric electrochemical signals in non-bilinear datasets or in the presence of irreversible electrochemical processes [68]. **Figure 6** summarizes the main steps of the fitting procedure.

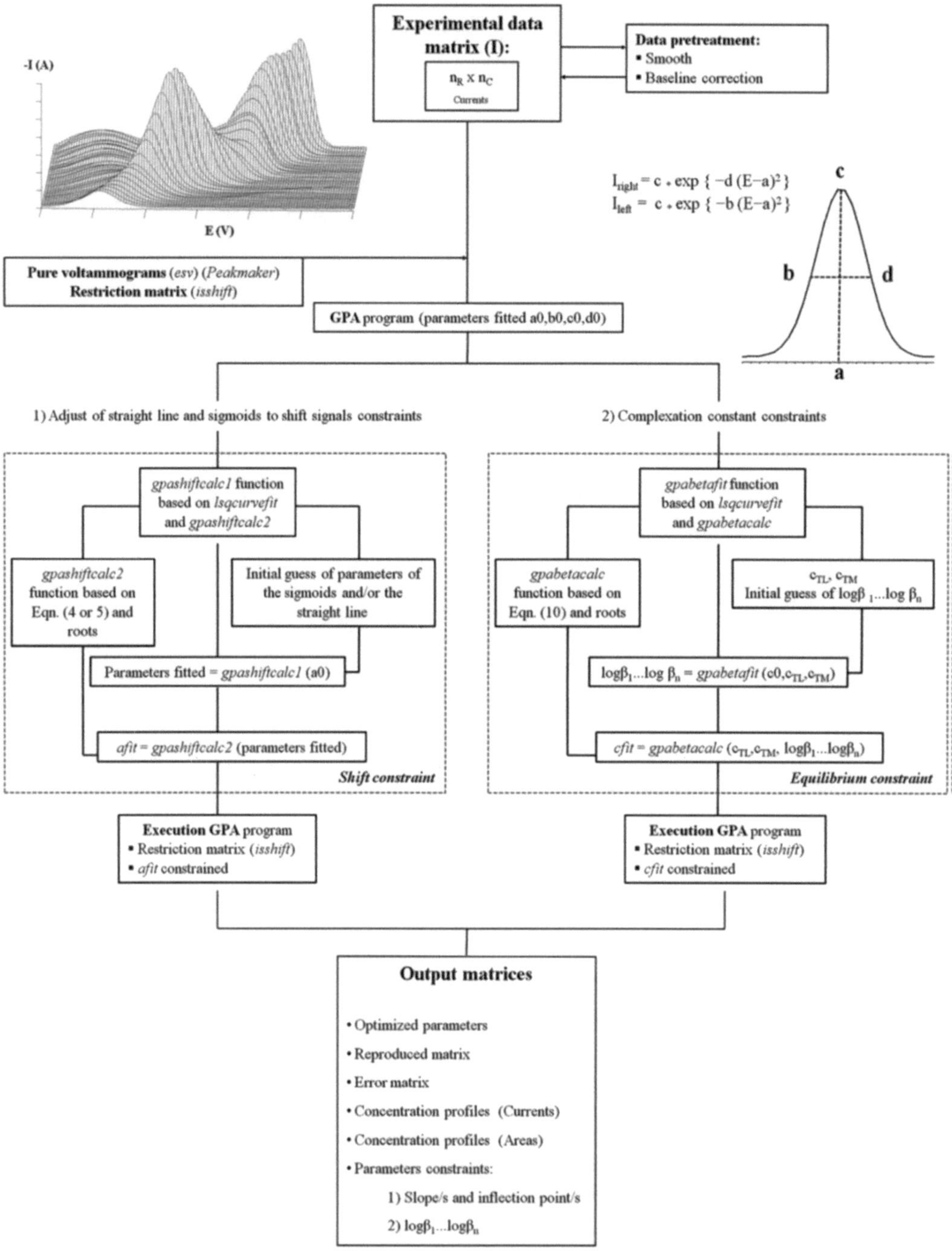

Figure 5. Flowchart of the GPA2D [68].

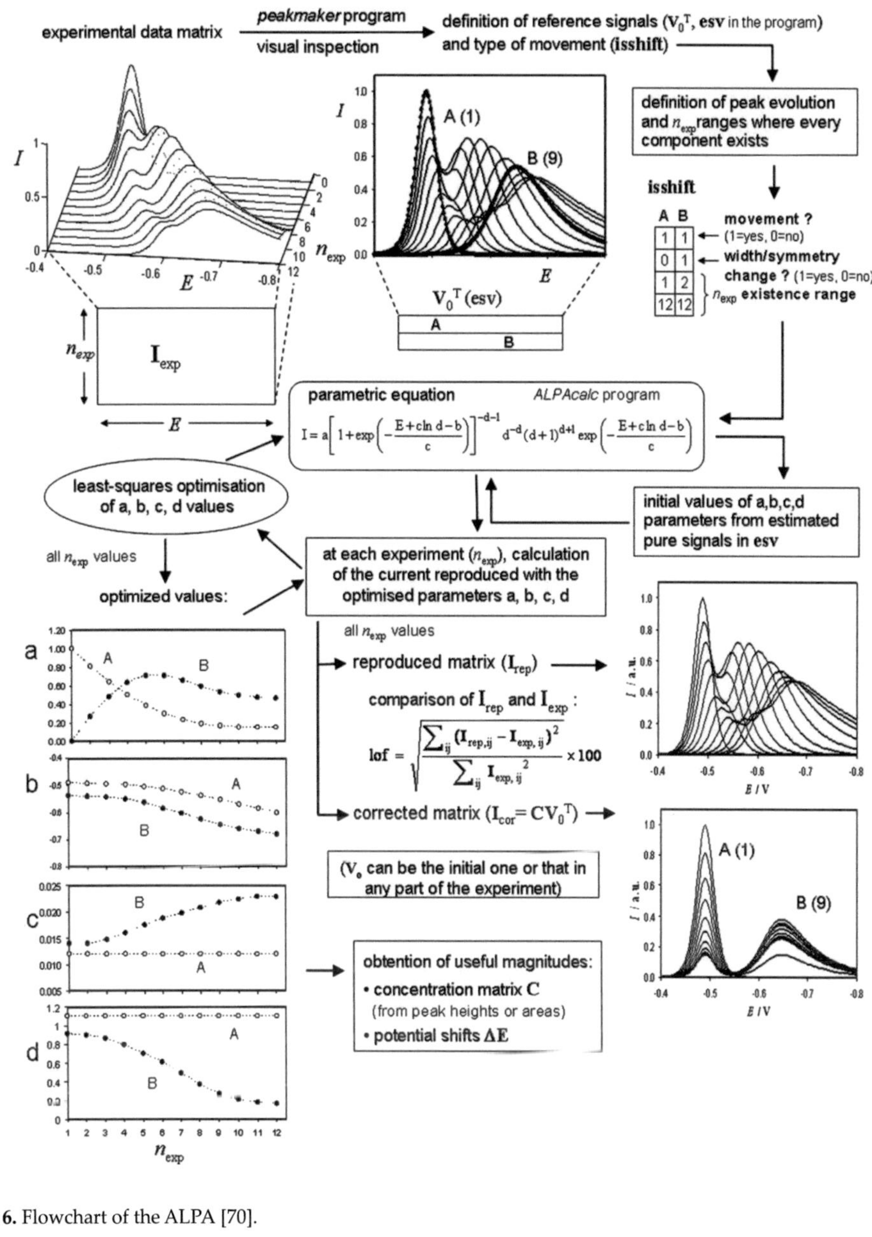

Figure 6. Flowchart of the ALPA [70].

1.8.2. Baseline correction

Baseline correction has been considered as a critical step for enhancing the signals and reducing the complexity of the analytical data [72, 73]. Considering this aim, Eilers et al. [74] have introduced an algorithm for baseline elimination based on asymmetric least squares splines regression (AsLSSR) approach. Details of the implementation of the mentioned method can be found in the literature [74, 75].

2. Applications of first-order multivariate calibration

Tables 1–3 summarize applications of first-order multivariate calibration with different first-order algorithms to electroanalytical data.

Technique	Application	Refs.
DPASV	Determination of Tl and Pb	[76]
PSA (at Au electrode)	Determination of As in the presence of Cu and Sn	[77]
DPP	Determination of furaltadone, furazolidone and nitrofurantoin	[78]
DPV	Determination of 2-(3)-t-butyl-4-methoxyphenol and propyl gallate	[79]
NPP, DPP	Determination of furazolidone and furaltadone	[80]
DPP	Determination of sulfadiazine, sulfamerazine and sulfamethazine	[81]
SWV, SWAdSV	Determination of sulphamethoxypyridazine and trimethoprim in veterinary formulations	[82]
DPP	Determination of Cu, Pb, Cd and Zn	[83]
LSV	Determination of indomethacin and acemethacin	[84]
ASV	Determination of Tl and Pb	[85]
DPP	Determination of Pb, Cd and Sn(IV)	[86]
LSV, CV, DC, DPP	Determination of propylgallate, butylated hydroxyanisole and butylated hydroxytoluene	[87]
DPASV	Determination of Cu in the presence of Fe	[88]
DPP	Variable selection for the determination of benzaldehyde and of Cu, Pb, Cd and Zn	[89]
DPASV	Variable selection for the determination of the binary mixtures Tl/Pb and Cu/Fe(III)	[90]
SWV, DPV	Determination of paraquat and diquat	[91]
DPAdSV	Speciation of Cr (determination of Cr(III) and Cr(VI))	[92, 94]
FIA-ED	Determination of 4-nitrophenol, phenol and p-cresol	[93]

http://dx.doi.org/10.5772/67310

Technique	Application	Refs.
DPV	Determination of the anti-inflammatory drugs indomethacin, acemethacin, piroxicam and tenoxicam	[95]
LSV	Determination of mixtures of vapours (ethanol, acetaldehyde, acetylene, SO_2, NO_2, NO, O_3)	[96]
DPAdSV	Determination of Al and Cr(VI)	[97]
DPV	Determination of nordihydroguaiaretic acid	[98]
DC, DPV, SWV	Determination of tocopherols in vegetable oils	[99]
NPP (at UME array)	Monitoring of *Staphylococcus aureus* population	[100]
NPP (at UME array)	Monitoring of *Escherichia coli* ATCC 13706 and *Pseudomonas aeruginosa* ATCC 27853 population	[101]
CV	Determination of cysteine, tyrosine and tryptophan	[102]
DPAdSV	Determination of In	[103]

Table 1. Application of partial least squares (PLS) method to electroanalytical data.

Technique	Application	Refs.
DPASV	Determination of Pb, Cd, Tl and In	[104]
PSA (at Au electrode)	Determination of Cu, Zn, Cd and Pb	[105]
DPP	Monitoring of freshness of milk (by an electronic tongue)	[106]
DPV	Determination of Cu	[107]
NPP, DPP	Determination of adenine and cytosine	[108]
DPP	Determination of Cu and Mo	[109]
SWV, SWAdSV	Determination of ethanol, fructose and glucose	[110]
DPP	Determination of vitamins B6 and B12 in	[111]
LSV	Determination of Cu	[112]
ASV	Determination of ethanol, methanol, fructose and glucose	[113]
DPP	Determination of nalidixic acid and its metabolite 7-hydroxymethylnalidixic acid	[114]

Table 2. Application of artificial neural networks (ANN) method to electroanalytical data.

Chemometrical technique	Electrochemical technique	Application	Refs.
MLR/PLS	DPASV	Study of influence of pH and Ca in metal/fulvic interactions	[115]
CLS/PLS/PCR	AdSV	Determination of synthetic colorants	[116]
CLS/ILS/KF	SWASV	Determination of Pb, Cd, In and Tl	[117]
CLS/PLS/PCR/MLR	DPP, NPP	Determination of Pb, Cd, Cu, Ni and V	[118]
CLS/PLS/PCR/MLR	ASV	Determination of Pb, Cd, Cu and Zn	[119]
PLS/NL-PLS/PCR	CV	Determination of tryptophan in feed samples	[120]
PLS/NL-PLS/PCR/MLR/ ANN	DuPSV	Determination of ethanol, fructose and glucose	[121]
PLS/PCR	CV	Determination of cysteine, tyrosine and tryptophan	[122]
PLS/ANN	DPV	Determination of catechol and hydroquinone at C fiber electrode	[123]
PLS/ANN	DC, DPP	Determination of atrazine/ simazine and terbutryn/ prometryn	[124]
CLS/PLS/PCR	LSV	Determination of synthetic food antioxidants	[125]
CLS/PLS/PCR/MLR	DPSV	Determination of chlorpromazine and promethazine hydrochloride	[126]
CLS/PLS/PCR/MLR	DPSV	Determination of five nitro-substituted aromatic compounds	[127]
PLS/PCR	ASV	Determination of Pb, Cd, In and Tl	[128]
CLS/PLS/MCR-ALS	ASV	Determination of Pb, Cd, In and Tl	[129]
PLS/PCR/ANN	ASV	Determination of Pb and Tl	[130]
PLS/PCR	DPSV	Determination of paracetamol and phenobarbital in pharmaceuticals	[131]
CLS/PCR/PLS/KF/ANN	DPSV	Determination of parathion, fenitrothion and parathion	[132]

http://dx.doi.org/10.5772/67310

Chemometrical technique	Electrochemical technique	Application	Refs.
PCR/PLS/GA-PLS/ANN	DPV	Determination of cysteine, tyrosine and tryptophan	[133]
CLS/PCR/PLS/ANN	DPV	Determination of propoxur, isoprocarb, carbaryl and carbofuran	[134]
HPCR/HPLS/CPCR/MBPLS	ACV	Determination of brightener in industrial Cu electroplating baths	[135]
PLS/ANN	CV	Determination of isoniazid and hydrazine	[136]

Table 3. Applications of different multivariate analysis techniques as applied to electroanalytical data.

3. Applications of second- and third-order multivariate calibrations

In an interesting work, Kooshki et al. generated three-way DPV data at different pulse heights of 20–100 mV with a 20 mV interval and analyzed them by MCR-ALS for determination of tryptophan (Trp) in the presence of tyrosine (Tyr) as an uncalibrated interference at the gold nanoparticles decorated multiwalled carbon nanotube modified glassy carbon electrode (Au NPs/GCE) [60]. The data were non-bilinear; therefore, the *shiftfit* algorithm was used to correct the observed shift in the data. **Figure 7** shows the potential shift correction of the augmented data for three standard Trp solutions (top left) and for a synthetic mixture solution containing Trp and Tyr (top right). These corrected data were augmented, and MCR-ALS was performed on them. The results of the potential shift correction and the MCR-ALS analysis for the determination of Trp in the synthetic mixtures confirmed that the analysis of the shift corrected data generates convergence with a low lack of fit value. Finally, they assessed the analytical utility of the proposed method by applying it to the determination of Trp in a fresh meat sample.

Galeano-Diaz et al. have reported a work based on adsorptive stripping square wave voltammetry (Ad-SSWV) for the simultaneous determination of fenitrothion (FEN) and its metabolites: fenitrooxon (OXON) and 3-methyl-4-nitrophenol (3-MET) in environmental samples [137]. These three compounds produced an electrochemical signal due to an adsorptive-reductive process at hanging mercury drop electrode (HMDE). The electrochemical approach showed a very high overlap degree for FEN and OXON voltammograms. Second-order multivariate calibration has been tested to solve the mixture of these three compounds. The accumulation time (t_{acc}) was chosen as the third variable (third way). The t_{acc} was varied in 5 s intervals and with the aim of increasing the total t_{acc} value without electrode saturation, the equilibration time was fixed at 5 s, and t_{acc} was varied from 5 to 25 s, thus five voltammograms for each sample were recorded. For the second-order multivariate calibration N-PLS/RBL, U-PLS/RBL and PARAFAC have been tested, using the three-way data intensity-potential-accumulation time. The U-PLS/RBL model was stated as the best second-order algorithm for the simultaneous determination of these three compounds. Finally, the proposed method was used to the analysis of river water samples as real cases and the results were encouraging.

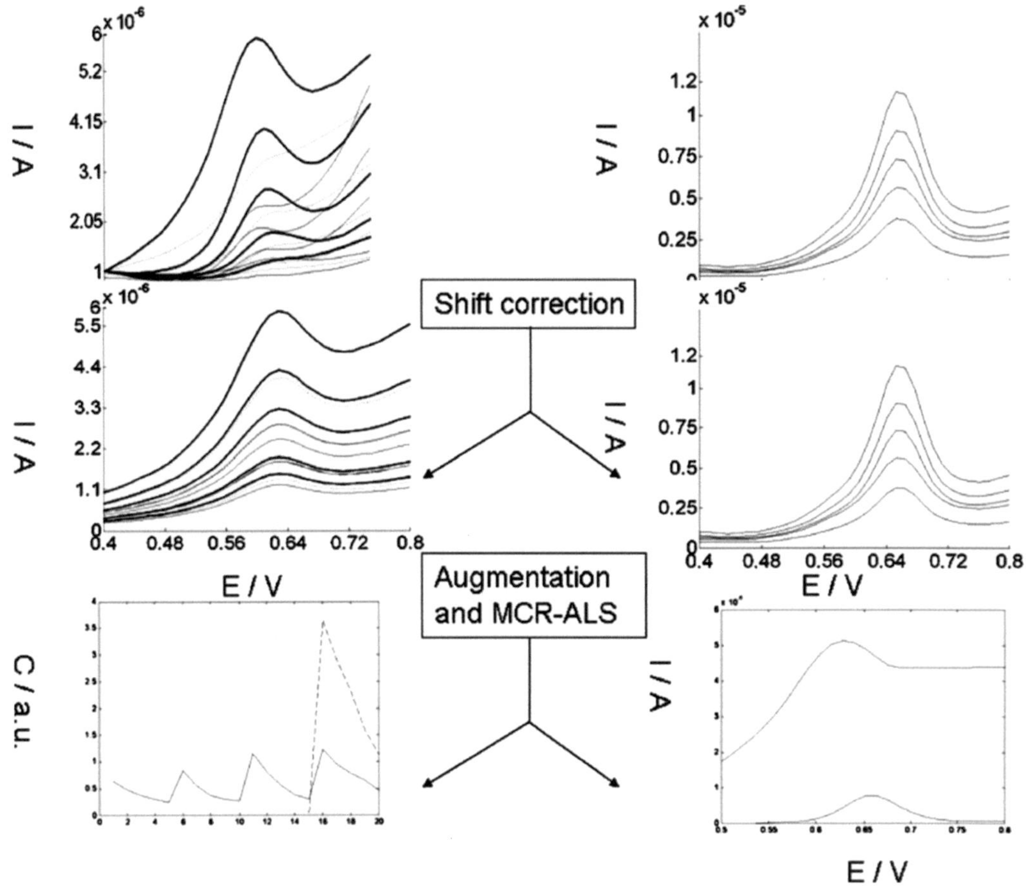

Figure 7. Application of potential shift correction, augmentation of the corrected data and resolution of the augmented data by MCR-ALS for the experimental data of standards (top left) and of a synthetic mixture (top right) [60].

Another interesting work entitled 'second-order data obtained from differential pulse voltammetry: determination of lead in river water using multivariate curve resolution-alternating least-squares (MCR-ALS)' was reported by Abdollahi et al. [24]. In this work, the MCR-ALS has been applied to potential-time second-order data with the aim of achieving the electrochemical second-order advantage. A simple way (change in pulse duration) was reported as the first approach towards generation of second-order DPV data. A linear dependency exists in the pulse duration profiles of the electroactive species in the mixture samples. Rank deficiency of the mixture data matrix was broken by matrix augmentation. Due to existence of potential shift in the obtained data, MCR-ALS could not be achieved the convergence on the augmented data. So, this shift was corrected with *shiftfit* program. Results of MCR-ALS after shift correction show that the proposed method could be efficiently used for determination of Pb^{2+} in the presence of unexpected interferents in the river water sample.

Khoobi et al. coupled DPV with MCR-ALS for simultaneous determination of betaxolol (Bet) and atenolol (Ate) at a multi-walled carbon nanotube modified carbon paste electrode (MWCNT/CPE) [138]. Operating conditions were optimized with central composite rotatable design (CCRD) and response surface methodology (RSM). Then, the second-order DPV data were generated at different pulse heights and after potential shifts correction by COW algorithm were analysed by MCR-ALS. **Figure 8** shows the resolved voltammograms of the seven mixtures of Bet and Ate that were applied in calibration curve, after using COW and MCR-ALS procedures. Finally, the developed method was successfully applied to simultaneous determination of Bet and Ate in human plasma.

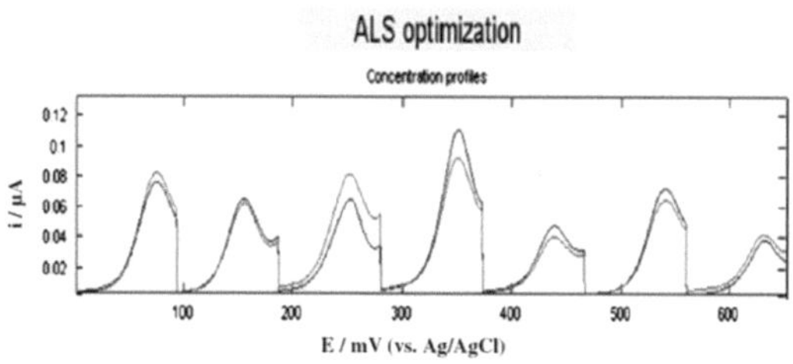

Figure 8. Application of potential shift correction using COW, augmentation of the corrected data and resolving of the augmented data by MCR-ALS for the data of seven standards of Bet and Ate [138].

In an interesting work by Khoobi et al. the MCR-ALS was used for determination of dopamine (DA) in the presence of epinephrine (EP) using second-order DPV data at different pulse heights on a carbon paste electrode modified with gold nanoparticles (AuNPs/CPE) [139]. The CCRD was employed to generate an experimental programme to model the effects of different parameters on voltammetric responses and the RSM was applied to show the individual and interactive effects of variables on the responses. The voltammograms of the samples were then collected into a column-wise augmented data matrix and subsequently analyzed by MCR-ALS. The effect of rotational ambiguity associated with a particular MCR-ALS solution under a set of constraints was also studied. With the aid of MCR-BANDS method, the absence of rotational ambiguity was verified. Finally, by the developed methodology, satisfactory results were obtained for the determination of DA in the presence of EP in spiked human blood plasma samples.

Ghoreishi et al. have reported a work based on coupling of three-way calibration with second-order DPV data for simultaneous quantification of sulfamethizole (SMT) and sulfapyridine (SPY) [140]. After finding the optimized values of the variables which affected the voltammetric responses, potential shift corrected by COW was used for further processing by MCR-ALS. Finally, the method was applied for simultaneous determination of SMT and SPY in spiked human serum and urine samples.

Masoum et al. generated second-order electrochemical data by changing the pulse height as an instrumental parameter [141]. After potential shift correction, MCR-ALS results showed that second-order calibration could be applied with great success for (+)-catechin determination in the presence of gallic acid at the surface of the multi-walled carbon nanotubes modified carbon paste electrode. The ability of the proposed method was evaluated using (+)-catechin determination in the presence of gallic acid in a green tea sample. In this study, fixed size moving window-evolving factor analysis (FSMW-EFA) [142] was used for the determination of pure variables, zero concentration and selective regions. Result of FSMW-EFA is shown in **Figure 9**. There are two curves higher than the noise level at peak region in FSWM-EFA plot. In this plot, regions that do not have any curves higher than the noise level are zero regions; regions that have one curve higher than the noise level are pure region (part 1 for gallic acid and part 2 for (+)-catechin) and regions that have two or more curves higher than the noise level are overlapped region (part 1 + 2). As shown in **Figure 9**, both components have selective region, so selectivity constraint can be applied. The solution to the problem of rank deficiency was the combined analysis of the rank deficient matrix with other matrices in the column direction that can have the suitable information to detect the presence of the hidden components [143].

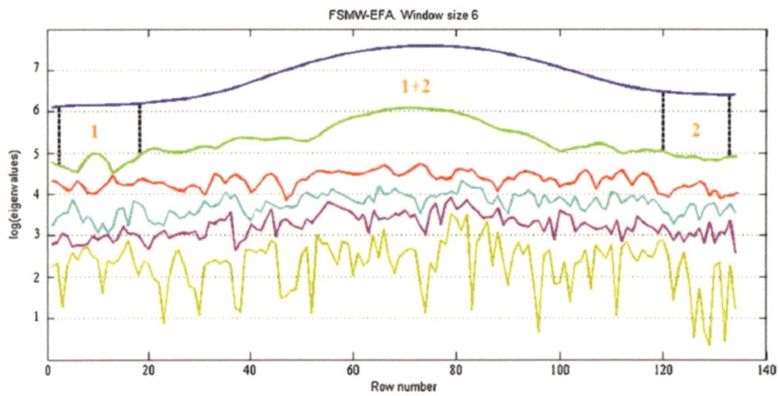

Figure 9. FSWM-EFA plot of real sample [141].

A work entitled 'application of Fe doped ZnO nanorods-based modified sensor for determination of sulfamethoxazole (SMX) and sulfamethizole (SMT) using chemometric methods in voltammetric studies' has been reported by Meshki et al. [144]. In this work, the second-order DPV data have been produced by changing the pulse heights and after potential shift correction with the help of COW algorithm they further processed by MCR-ALS for exploiting second-order advantage. The potential shift correction was carried out on a column-wise augmented data matrix that contained 13 calibration set of SMX and SMT. Then MCR-ALS was performed on the new augmented data and lack of fit was reduced and was better than that obtained in the absence of potential shift correction. Finally, the application of the proposed method was examined for simultaneous determination of SMX and SMT in human blood serum and urine samples.

Jalalvand et al. have reported a work for generation of second-order DPV data based on changing the pulse heights and application of them for simultaneous quantification of norepinephrine (NE), paracetamol (AC) and uric acid (UA) in the presence of pteroylglutamic acid (FA) as an uncalibrated interference at an electrochemically oxidized glassy carbon electrode (OGCE) [145]. In this work, several second-order calibration models based on ANN-RBL, U-PLS/RBL, N-PLS/RBL, MCR-ALS and PARAFAC2 were used to exploiting second-order

http://dx.doi.org/10.5772/67310

advantage to identify which technique offers the best predictions. The baseline of the DPV signals was corrected by asymmetric least squares spline regression (AsLSSR) algorithm and the observed shifts were corrected using COW algorithm. All the algorithms achieved the second-order advantage and were in principle able to overcome the problem of the presence of unexpected interference. Comparison of the performance of the applied second-order chemometric algorithms confirmed the more superiority of U-PLS/RBL to resolve complex systems (see **Figure 10**). The results of applying U-PLS/RBL for the simultaneous quantification of the studied analytes in human serum samples were also encouraging.

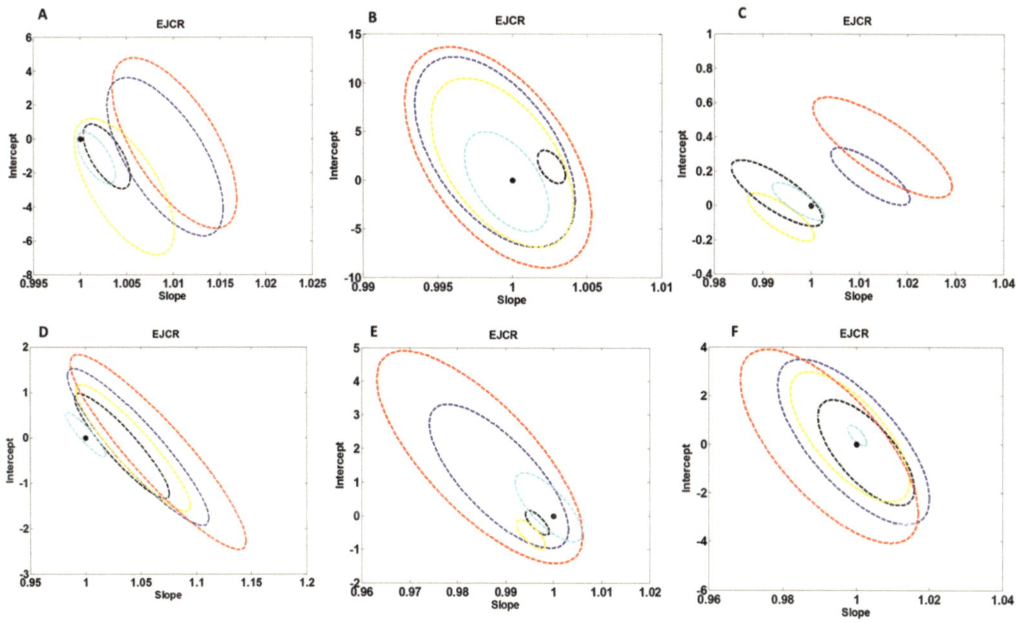

Figure 10. Elliptical joint regions (at 95% confidence level) for the slopes and intercepts of the regressions for (A) AC, validation set, (B) AC, test set, (C) NE, validation set, (D) NE, test set, (E) UA, validation set and (F) UA, test set. In all cases: black point marks the theoretical point (0,1), black ellipse shows ANN results, blue ellipse shows MCR-ALS results, yellow ellipse shows N-PLS/RBL results, red ellipse shows PARAFAC2 results and cyan ellipse shows U-PLS/RBL results [145].

Mora Diez et al. have reported the work to develop a method based on DPV coupled to second-order data modelling with MCR-ALS and U-PLS/RBL for the quantitation of the pesticide ethiofencarb in the presence of fenobucarb and bendiocarb as interferences in tap water [146]. In this study, the possibility of second-order multivariate calibration was studied by using the hydrolysis time as the third variable, and MCR-ALS and U-PLS/RBL. Asymmetric least squares background correction adapted to second-order data was used to remove the baseline of the data (**Figure 11A**). **Figure 11B** shows the voltammograms retrieved by MCR-ALS for all the three mentioned components in validation sample number 1. As can be appreciated there exists a high degree of overlapping among analyte and interferents signals. In addition, **Figure 11C** shows the corresponding time evolution profiles in this particular sample and the

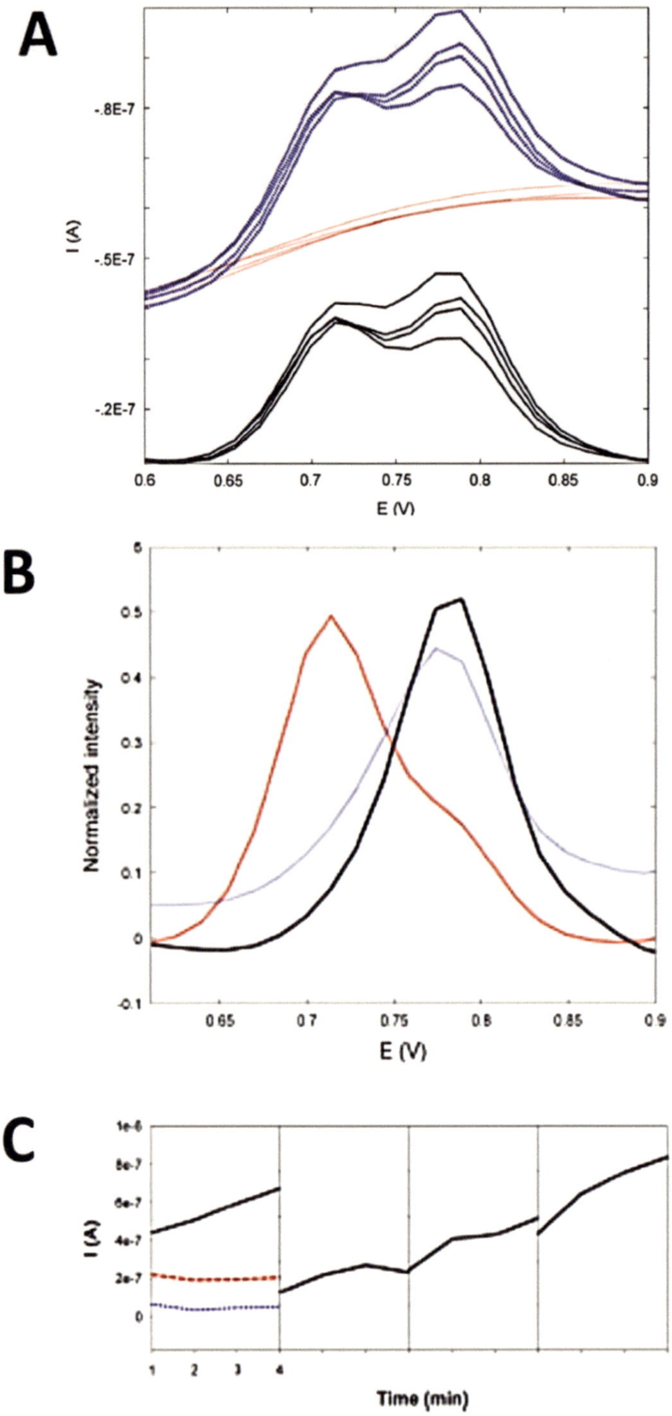

Figure 11. (A) Voltammograms corresponding to validation sample number 1, before and after background correction. (B) Voltammogram profiles retrieved with MCR-ALS for validation sample number 1 and (C) time profiles retrieved with MCR-ALS when analysing validation sample number 1 and three pure standard solutions of ethiofencarb [146].

http://dx.doi.org/10.5772/67310

ones retrieved for three ethiofencarb standard samples. The areas under the kinetic profiles were used to build a calibration curve that allowed them to obtain the concentration of ethiofencarb in the validation samples. After model building by U-PLS/RBL, the outputs of MCR-ALS and U-PLS/RBL were compared by elliptical joint confidence region (EJCR) method and EJCR confirms the better performance of U-PLS/RBL than MCR-ALS.

Granero et al. have reported a work based on three-way calibration with second-order square wave voltammetric (SWV) data for simultaneous determination of ascorbic acid, uric acid, and dopamine in the presence of glucose (interfering species) in lyophilized human serum samples [147]. The second-order data were baseline- and shift-corrected by AsLSSR and COW algorithms, respectively, and then modelled by U-PLS/RBL second-order algorithm. Finally, the developed analytical method was successfully applied to determine ascorbic acid, uric acid, and dopamine in lyophilized human serum samples.

Jaworski et al. have reported a work related to the application of multi-way chemometric techniques for the analysis of AC voltammetric data [148]. In this study, three multi-way calibration techniques have been applied for determining the suppressor concentration in industrial copper electrometallization baths used in semiconductor manufacturing. PARAFAC for multi-way array decomposition coupled with inverse least squares (ILS) regression (PARAFAC/ILS), DTLD coupled with ILS (DTLD/ILS), and multilinear partial least squares (NPLS) regression were employed to develop and test calibration models based on trilinear AC voltammetric data. The hardships associated with the physical interpretation of very complex AC voltammograms were tackled by the use of powerful chemometric tools which played a significant role in the revival of interest in real-life applications of AC based electroanalytical techniques.

Recently, an interesting work has been published by Jalalvand et al., which reports coupling of four-way multivariate calibration with third-order DPV data [62]. To achieve this goal, the DPV response of each sample was recorded 36 times. Six current-potential matrices were recorded at six different pulse durations. Each matrix consists of six vectors which have been recorded at six different pulse heights. The three-way data array obtained for the calibration set and for each of the test samples were joined into a single four-way data array. The recorded data were baseline-corrected by AsLSSR and the data array was nonlinear, thus, the non-linearities were tackled by potential shift correction using COW algorithm (see **Figure 12**) and subsequently was analysed with U-PLS/RTL and N-PLS/RTL as third-order multivariate calibration algorithms. A comprehensive and systematic strategy for comparing the performance of the two algorithms was presented in this work, in particular with a view of practical applications. This comparison was developed to identify which algorithm offers the best predictions for the simultaneous determination of levodopa (LD), carbidopa (CD), methyldopa (MD), acetaminophen (AC), tramadol (TRA), lidocaine (LC), tolperisone (TOP), ofloxacin (OF), levofloxacin (LOF) and norfloxacin (NOF) in the presence of benserazide (BA), dopamine (DP) and ciprofloxacin (COF) as uncalibrated interferences using a multi-walled carbon nanotubes modified glassy carbon electrode (MWCNTs/GCE). This study demonstrated the more superiority of U-PLS/RTL to resolve the complex systems. The results of applying U-PLS/RTL for the simultaneous determination of the studied analytes in human serum samples as experimental cases were also encouraging.

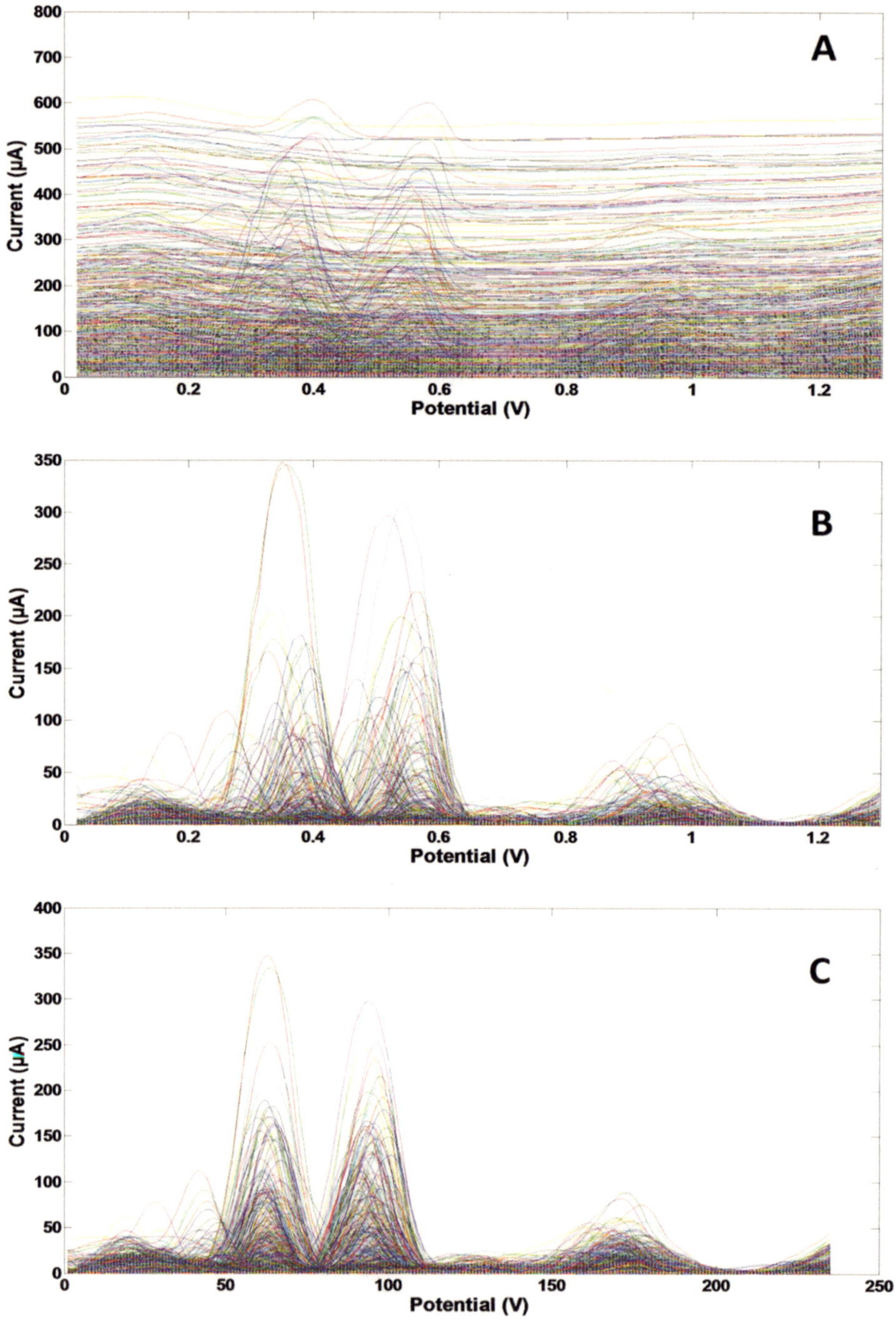

Figure 12. Differential pulse voltammetric data corresponding to the calibration set. (A) Raw data, and after preprocessing: (B) baseline correction and (C) alignment with COW [62].

http://dx.doi.org/10.5772/67310

4. Conclusions

Multi-dimensional data are being abundantly produced by modern analytical instrumentation, calling for new and powerful data-processing techniques. Research in the last two decades has resulted in the development of a multitude of different processing algorithms, each equipped with its own sophisticated artillery. Going from univariate data (a single datum per sample, employed in the well-known classical univariate calibration) to multivariate data (data arrays per sample of increasingly complex structure and number of dimensions) is known to provide a gain in sensitivity and selectivity, combined with analytical advantages which cannot be overestimated. Nowadays, chemometrics is essential to exploiting the extraordinary potential of modern analytical instruments. This has been widely demonstrated with different types of signals. Electroanalytical chemistry cannot ignore this dominant trend. Modern electrochemical instrumentation provides reliable and reproducible data that are the basis of analytical methods with very low quantitation limits. Electrochemical methods are very interesting techniques for coupling with multi-way calibration because they provide excellent and low-cost opportunities for accurate and reliable determination of analyte(s) and because of the existence of instrumental parameters; they are very suitable for generating second- and third-order data. The second-order advantage, achieved with second- (or higher-) order sample data, allows one not only to mark new samples containing components which do not occur in the calibration phase but also to model their contribution to the overall signal, and most importantly, to accurately quantitate the calibrated analyte(s). Voltammetric measurements assisted by multi-way calibration are producing increasingly complex data structures, whose appropriate chemometric processing opens new dimensions in analytical studies. Improved sensitivity and selectivity, the possibility of analyte quantitation in the presence of uncalibrated interferents, and the possibility of obtaining qualitative interpretation of chemical phenomena through the study of multi-way data, in a much better way than with univariate or first-order data, are some of the advantages which can be achieved. The most problems with voltammetric data for coupling with multi-way calibration are the baseline of the signals and sample-to-sample potential shifts in the analyte profiles and for tackling these problems chemometric tools can show an interesting power. The chemometric algorithms such as COW, icoshift, *shiftfit*, *pHfit*, GPA, GPA2D and ALPA can be used for correcting the shifts and the baseline of the signals could be removed with AsLSSR as a power chemometric tool. On the whole, voltammetric measurements assisted by multi-way calibration are gaining attention of the scientists and we hope this review will help to promote the use of multi-way calibration in electroanalytical chemistry.

Acknowledgements

ARJ wishes this chapter to be useful for the electrochemists to promote the use of chemometrics in electrochemistry.

Author details

Ali R. Jalalvand

Address all correspondence to: ali.jalalvand1984@gmail.com

Department of Chemical Engineering, Faculty of Energy, Kermanshah University of Technology, Kermanshah, Iran

References

[1] D.L. Massart, B.G.M. Vandeginste, L.M.C. Buydens, S. De Jong, P.J. Lewi, J. Smeyers-Verbeke, Handbook of chemometrics and qualimetrics Part A, Amsterdam, Elsevier, 1998.

[2] G.M. Escandar, N.M. Faber, H.C. Goicoechea, A. Munoz de la Pena, A.C. Olivieri, R.J. Poppi, Trends Anal. Chem. 26 (2007) 752–765.

[3] R. Bro, Crit. Rev. Anal. Chem. 36 (2006) 279–293.

[4] A.C. Olivieri, Anal. Chem. 80 (2008) 5713–5720.

[5] K.S. Booksh, B.R. Kowalski, Anal. Chem. 66 (1994) 782–791.

[6] J.C. Wang, Analytical electrochemistry, Chichester, Wiley, 2000.

[7] H.H. Girault, Analytical and physical electrochemistry, Lausanne, EPFL, 2004.

[8] K.I. Ozomwna, Recent advances in analytical electrochemistry, India, Transworld Research Network, 2007.

[9] E.A.M.F. Dahmen, Electroanalysis: theory and applications in aqueous and non-aqueous media and in automated chemical control, Amsterdam, Elsevier, 1986.

[10] M. Esteban, C. Arino, J.M. Diaz-Cruz, Crit. Rev. Anal. Chem. 36 (2006) 295–313.

[11] S.D. Brown, R.S. Bear, Crit. Rev. Anal. Chem. 24 (1993) 99–131.

[12] Y. Ni, S. Kokot, Anal. Chim. Acta (2008) 130–146.

[13] V. Gomez, M.P. Callao, Anal. Chim. Acta (2008) 169–183.

[14] T. Madrakian, A. Afkhami, M. Mohammadnejad, Anal. Chim. Acta 645 (2009) 25–29.

[15] Y. Zhang, H.L. Wu, A.L. Xia, Q.J. Han, H. Cui, R.Q. Yu, Talanta 72 (2007) 926–931.

[16] A. Edelmann, J. Diewok, J.R. Baena, B. Lendl, Anal. Bioanal. Chem. 376 (2003) 92–97.

[17] E. Pere-Trepat, R. Tauler, J. Chromatogr. A 1131 (2006) 85–96.

[18] A. Checa, R. Oliver, J. Saurina, S. Hernandez-Cassou, Anal. Chim. Acta 572 (2006) 155–164.

[19] V. Río, M.S. Larrechi, M.P. Callao, Anal. Chim. Acta 676 (2010) 28–33.

[20] H.C. Goicoechea, M.J. Culzoni, M.D.G. García, M. Martínez Galera, Talanta 83 (2011) 1098–1107.

[21] M.J. Culzoni, P.C. Damiani, A. Garcia-Reiriz, H.C. Goicoechea, A.C. Olivieri, Analyst 132 (2007) 654–663.

[22] H.C. Goicoechea, A.C. Olivieri, Appl. Spectrosc. 59 (2005) 926–933.

[23] N.E. Llamas, M. Garrido, M.S.D. Nezio, B.S.F. Band, Anal. Chim. Acta 655 (2009) 38–42.

[24] H. Abdollahi, M. Kooshki, Electroanalysis 22 (2010) 2245–2253.

[25] https://en.wikipedia.org/wiki/Calibration.

[26] A.C. Olivieri, G.M. Escandar, Practical three-way calibration, Amsterdam, Elsevier, 2014.

[27] K. Danzer, L.A. Currie, Pure Appl. Chem. 70 (1998) 993–1014.

[28] D.A. Burns, E.W. Ciurczak, Handbook of near-infrared analysis in practical spectroscopy series, Boca Raton, FL, USA, CRC Press, 2008.

[29] H. Martens, T. Næs, Multivariate calibration, Chichester, Wiley, 1989.

[30] A. Smilde, R. Bro, P. Geladi, Multiway analysis with applications in the chemical sciences, West Sussex, England, Wiley, 2004.

[31] A.C. Olivieri, G.M. Escandar, A. Munoz de la Pena, Trends Anal. Chem. 30 (2011) 607–617.

[32] R. Bro, Chemom. Intell. Lab. Syst. 38 (1997) 149–171.

[33] E. Sanchez, B.R. Kowalski, Anal. Chem. 58 (1986) 496–499.

[34] E. Sanchez, B.R. Kowalski, J. Chemometrics 4 (1990) 29–45.

[35] A. De Juan, E. Casassas, R. Tauler, in: R.A. Myers (Editor), Encyclopedia of analytical chemistry, Vol. 11, Chichester, West Sussex, UK, Wiley, 2002.

[36] M. Linder, R. Sundberg, Chemom. Intell. Lab. Syst. 42 (1998) 159–178.

[37] M. Linder, R. Sundberg, J. Chemometrics 16 (2002) 12–27.

[38] H.L. Wu, M. Shibukawa, K. Oguma, J. Chemometrics 12 (1998) 1–26.

[39] Z.P. Chen, H.L. Wu, J.H. Jiang, Y. Li, R.Q. Yu, Chemom. Intell. Lab. Syst. 52 (2000) 75–86.

[40] L.Q. Hu, H.L. Wu, Y.J. Ding, D.M. Fang, A.L. Xia, R.Q. Yu, Chemom. Intell. Lab. Syst. 82 (2006) 145–153.

[41] A.L. Xia, H.L. Wu, D.M. Fang, Y.J. Ding, L.Q. Hu, R.Q. Yu, J. Chemometrics 19 (2005) 65–76.

[42] S. Wold, P. Geladi, K. Esbensen, J. Øhman, J. Chemometrics 1 (1987) 41–56.

[43] R. Bro, J. Chemometrics 10 (1996) 47–62.

[44] J. Ohman, P. Geladi, S. Wold, J. Chemometrics 4 (1990) 79–90.

[45] A.C. Olivieri, J. Chemometrics 19 (2005) 253–265.

[46] J.A. Arancibia, A.C. Olivieri, D. Bohoyo Gil, A. Espinosa Mansilla, I. Duran Meras, A. Munoz de la Pena, Chemom. Intell. Lab. System 80 (2006) 77–86.

[47] A.L. Xia, H.L. Wu, S.F. Li, S.H. Zhu, L.Q. Hu, R.Q. Yu, J. Chemometrics 21 (2007) 133.

[48] H.A.L. Kiers, J.M.F. Ten Berge, R. Bro, J. Chemometrics 13 (1999) 275–294.

[49] M. Bahram, R. Bro, Anal. Chim. Acta 584 (2007) 397–402.

[50] R. Tauler, Chemom. Intell. Lab. Syst. 30 (1995) 133–146.

[51] M.M. Reis, S.P. Gurden, A.K. Smilde, M.M.C. Ferreira, Anal. Chim. Acta 422 (2000) 21–36.

[52] R. Bro, J. Chemometrics 10 (1996) 47–61.

[53] T. Czekaj, W. Wu, B. Walczak, J. Chemometrics 19 (2005) 341–354.

[54] S. Haykin, Neural networks. A comprehensive foundation, 2nd edition, Upper Saddle River, NJ, USA, Prentice-Hall, 1999.

[55] F. Marini, R. Bucci, A.L. Magri, A.D. Magri, Microchem. J. 88 (2008) 178–185.

[56] V.A. Lozano, G.A. Ibanez, A.C. Olivieri, Anal. Chim. Acta 610 (2008) 186–195.

[57] M.J. Culzoni, H.C. Goicoechea, A.P. Pagani, M.A. Cabezon, A.C. Olivieri, Analyst 131 (2006) 718–723.

[58] A. Garcıa Reiriz, P.C. Damiani, M.J. Culzoni, H.C. Goicoechea, A.C. Olivieri, Chemom. Intell. Lab. Syst. 92 (2008) 61–70.

[59] A. Garcıa Reiriz, P.C. Damiani, A.C. Olivieri, Chemom. Intell. Lab. Syst. 100 (2010) 127–135.

[60] M. Kooshki, H. Abdollahi, S. Bozorgzadeh, B. Haghighi, Electrochim. Acta 56 (2011) 8618–8624.

[61] A.J. Bard, L.R. Faulkner, Electrochemical methods: fundamentals and applications, New York, John Wiley & Sons, Inc., 2001.

[62] A.R. Jalalvand, M.B. Gholivand, H.C. Goicoechea, Chemom. Intell. Lab. Syst. 148 (2015) 60–71.

[63] J. Listgarten, A. Emili, Mol. Cell. Proteomics 4 (2005) 419–434.

[64] G. Tomasi, F. Savorani, S.B. Engelsen, J. Chromatogr. A, 1218 (2011) 7832–7840.

[65] F. Savorani, G. Tomasi, S.B. Engelsen, J. Magn. Reson. 202 (2010) 190–202.

[66] S. Cavanillas, J.M. Diaz-Cruz, C. Arino, M. Esteban, Anal. Chim. Acta 689 (2011) 198–205.

[67] S. Cavanillas, N. Serrano, J.M. Diaz-Cruz, C. Arino, M. Esteban, Analyst 138 (2013) 2171–2180.

[68] M. Kooshki, J.M. Diaz-Cruz, H. Abdollahi, C. Arino, M. Esteban, Analyst 136 (2011) 4696–4703.

[69] A. Alberich, J.M. Diaz-Cruz, C. Arino, M. Esteban, Analyst 133 (2008) 112–125.

[70] A. Alberich, J.M. Diaz-Cruz, C. Arino, M. Esteban, Analyst 133 (2008) 470–477.

[71] J.M. Diaz-Cruz, J. Sanchis, E. Chekmeneva, C. Arino, M. Esteban, Analyst 135 (2010) 1653–1662.

[72] A.R. Jalalvand, M.B. Gholivand, H.C. Goicoechea, Å. Rinnan, T. Skov, Chemom. Intell. Lab. Syst. 146 (2015) 437–446.

[73] M.M. De Zan, M.D. Gil García, M.J. Culzoni, R.G. Siano, H.C. Goicoechea, M. Martínez Galera, J. Chromatogr. A 1179 (2008) 106–114.

[74] P.H.C. Eilers, I.D. Currie, M. Durban, Comput. Statist. Data Anal. 50 (2006) 61–76.

[75] P.H.C. Eilers, Anal. Chem. 76 (2004) 404–411.

[76] A. Henrion, R. Henrion, G. Henrion, F. Scholz, Electroanalysis 2 (1990) 309–312.

[77] D. Jagner, L. Renman, S.H. Stefansdottir, Electroanalysis 6 (1994) 201–208.

[78] A. Guiberteau, T. Galeano, A. Espinosa-Mansilla, F. Salinas, Talanta 41 (1994) 1821–1832.

[79] E. Martin, J.M. Garcıa, A.I. Jimenez, J.J. Arias, Quım. Anal. 14 (1995) 218–222.

[80] A. Guiberteau, T. Galeano, A. Espinosa-Mansilla, P.L. Lopez-deAlba, F. Salinas, Anal. Chim. Acta 302 (1995) 9–19.

[81] T. Galeano, A. Guiberteau, M.I. Acedo, F. Salinas, Analyst 121 (1996) 547–552.

[82] J.J. Berzas, J. Rodrıguez, G. Castaneda, Anal. Chim. Acta 349 (1997) 303–311.

[83] A. Herrero, M.C. Ortiz, Anal. Chim. Acta 348 (1997) 51 59.

[84] M.J. Arcos, C. Alonso, M.C. Ortiz, Electrochim. Acta 43 (1998) 479–485.

[85] A. Herrero, M.C. Ortiz, Talanta 46 (1998) 129–138.

[86] A. Herrero, M. C. Ortiz, Electroanalysis 10 (1998) 717–721.

[87] T. Galeano, A. Guiberteau, M.F. Alexandre, F. Salinas, J.C. Vire, Electroanalysis 10 (1998) 497–505.

[88] A. Herrero, M.C. Ortiz, Talanta 49 (1999) 801–811.

[89] A. Herrero, M.C. Ortiz, Anal. Chim. Acta 378 (1999)245–259.

[90] T. Galeano, A. Guiberteau, F. Salinas, Electroanalysis 12 (2000) 616–621.

[91] O. Domınguez, M.J. Arcos, Electroanalysis 12 (2000) 449–458.

[92] M.E. Rueda, L.A. Sarabia, A. Herrero, M.C. Ortiz, Anal. Chim. Acta 446 (2001) 269–279.

[93] O. Domınguez, M.J. Arcos, Anal. Chim. Acta 470 (2002) 241–252.

[94] C. Reguera, M.C. Ortiz, M.J. Arcos, Electroanalysis 14 (2002) 1699–1706.

[95] R. Knake, R. Guchardi, P.C. Hauser, Anal. Chim. Acta 475 (2003) 17–25.

[96] M. Camara, O. Domınguez, M.J. Arcos, Helv. Chim. Acta 86 (2003) 2434–2440.

[97] T. Galeano, A. Espinosa-Mansilla, B. Roldan, F. Salinas, Electroanalysis 15 (2003) 646–651.

[98] T. Galeano, I. Duran, A. Guiberteau, M. F. Alexandre, Anal. Chim. Acta 511 (2004) 231–238.

[99] M. Berrettoni, D. Tonelli, P. Conti, R. Marassi, M. Trevisani, Sens. Actuators, B 102 (2004) 331–335.

[100] M. Berrettoni, I. Carpani, N. Corradini, P. Conti, G. Fumarola, G. Legnani, S. Lanteri, R. Marassi, D. Tonelli, Anal. Chim. Acta 509 (2004) 95–101.

[101] L. Moreno, A. Merkoci, S. Alegret, S. Hernandez-Cassou, J. Saurina, Anal. Chim. Acta 507 (2004) 247–253.

[102] I. Paolicchi, O. Domınguez, M. A. Alonso, M. J. Arcos, Anal. Chim. Acta 511 (2004) 223–229.

[103] F. Despagne, D. L. Massart, Analyst 123 (1998) 157–178.

[104] H. Chan, A. Butler, D.M. Falck, M. S. Freund, Anal. Chem. 69 (1997) 2373–2378.

[105] F. Winquist, C. Krantz-Rulcker, P. Wide, I. Lundstrom, Meas. Sci. Technol. 9 (1998) 1937–1946.

[106] T. Khayamian, A.A. Ensafi, M. Atabati, Microchem. J. 65 (2000) 347–351.

[107] E. Cukrowska, L. Trnkova, R. Kizek, J. Havel, J. Electroanal. Chem. 503 (2001) 117–124.

[108] A.A. Ensafi, T. Khayamian, M. Atabati, Talanta 57 (2002) 785–793.

[109] C. Bessant, S. Saini, Anal. Chem. 71 (1999) 2806–2813.

[110] S.R. Hernandez, G.G. Ribero, H.C. Goicoechea, Talanta 61 (2003) 743–753.

[111] A.A. Ensafi, T. Khayamian, A. Benvidi, Can. J. Anal. Sci. Spectrosc. 49 (2004) 271–276.

[112] E. Richards, C. Bessant, S. Saini, Analyst 129 (2004) 355–358.

[113] A. Guiberteau, T. Galeano, M. Rodrıguez, J. M. Ortiz, I. Duran, F. Salinas, Talanta 62 (2004) 357–365.

[114] W. von Tuempling, S. Geiss, J. Einax, Acta Hydrochim. Hydrobiol. 20 (1992) 320–325.

[115] Y. Ni, J. Bai, L. Jin, Anal. Chim. Acta 329 (1996) 65–72.

[116] H.N.A. Hassan, M.E.M. Hassouna, I. H. I. Habib, Talanta 46 (1998) 1195–1203.

[117] Y. Ni, L. Jin, Chemom. Intell. Lab. Syst. 45 (1999) 105–111.

[118] A. Donachie, A.D. Walmsley, S.J. Haswel, Anal. Chim. Acta 378 (1999) 235–243.

[119] J. Saurina, S. Hernandez-Cassou, E. Fabregas, S. Alegret, Analyst 124 (1999) 733–737.

[120] C. Bessant, S. Saini, J. Electroanal. Chem. 489 (2000) 76–83.

[121] J. Saurina, S. Hernandez-Cassou, E. Fabregas, S. Alegret, Anal. Chim. Acta 405 (2000) 153–160.

[122] R.M. de Carvalho, C. Mello, L.T. Kubota, Anal. Chim. Acta 420 (2000) 109–121.

[123] A. Guiberteau, T. Galeano, N.M. Mora, F. Salinas, J.M. Ortiz, J.C. Vire, Analyst 125 (2000) 909–914.

[124] Y. Ni, L. Wang, S. Kokot, Anal. Chim. Acta 412 (2000) 185–193.

[125] Y. Ni, L. Wang, S. Kokot, Anal. Chim. Acta 439 (2001) 159–168.

[126] Y. Ni, L. Wang, S. Kokot, Anal. Chim. Acta 431 (2001) 101–113.

[127] M.C. Antunes, J.E. Simao, A.C. Duarte, Electroanalysis 13 (2001) 1041–1045.

[128] M.C. Antunes, J.E. Simao, A.C. Duarte, Analyst 127 (2002) 809–817.

[129] J.M. Palacios, A. Jimenez, L.M. Cubillana, I. Naranjo, J.L. Hidalgo, Microchim. Acta 142 (2003) 27–36.

[130] Y. Ni, Y.Wang, S. Kokot, Anal. Lett. 37 (2004) 3219–3235.

[131] Y. Ni, P. Qui, S. Kokot, Anal. Chim. Acta 516 (2004) 7–17.

[132] M.R. Majidi, K. Asadpour-Zeynali, J. Chin. Chem. Soc. 52 (2005) 21–28.

[133] Y. Ni, P. Qui, S. Kokot, Anal. Chim. Acta 537 (2005) 321–330.

[134] A. Jaworski, K. Wikiel, H. Wikiel, Electroanalysis 17 (2005) 1477–1485.

[135] M.R. Majidi, A. Jouyban, K. Asadpour-Zeynali, Electroanalysis 17 (2005) 915–918.

[136] R. Barthus, L.H. Mazo, R.J. Poppi, J. Pharm. Biomed. Anal. 38 (2005) 94–99.

[137] T. Galeano-Dıaz, A. Guiberteau-Cabanillas, A. Espinosa-Mansilla, M.D. Lopez-Soto, Anal. Chim. Acta 618 (2008) 131–139.

[138] A. Khoobi, S.M. Ghoreishi, S. Masoum, M. Behpour, Bioelectrochemistry 94 (2013) 100–107.

[139] A. Khoobi, S.M. Ghoreishi, M. Behpour, S. Masoum, Anal. Chem. 86 (2014) 8967–8973.

[140] S.M. Ghoreishi, A. Khoobi, M. Behpour, S. Masoum, Electrochim. Acta 130 (2014) 271–278.

[141] S. Masoum, M. Behpour, F. Azimi, M.H. Motaghedifard, Sens. Actuat. B 193 (2014) 582–591.

[142] H. Keller, D. Massart, Anal. Chim. Acta 246 (1991) 379–390.

[143] A. de Juan, M. Maeder, M. Martinez, R. Tauler, Chemometr. Intell. Lab. 54 (2000)123–141.

[144] M. Meshki, M. Behpour, S. Masoum, J. Electroanal. Chem. 740 (2015) 1–7.

[145] A.R. Jalalvand, M.B. Gholivand, H.C. Goicoechea, T. Skov, Talanta 134 (2015) 607–618.

[146] N.M. Diez, A.G. Cabanillas, A.S. Rodríguez, H.C. Goicoechea, Talanta 132 (2015) 851–856.

[147] A.M. Granero, G.D. Pierini, S.N. Robledo, M. S. Di Nezio, H. Fernandez, M.A. Zon, Microchem. J. 129 (2016) 205–212.

[148] A. Jaworski, H. Wikiel, K. Wikiel, Electroanalysis 21 (2009) 580–589.

Chapter 2

Study of Metrological Properties of Voltammetric Electrodes in the Time Domain

Krzysztof Suchocki

Additional information is available at the end of the chapter

http://dx.doi.org/10.5772/67944

Abstract

Metrological properties of voltammetric electrodes, in the situation where on their surface an electrochemical reaction of oxidizing/reduction takes place, were analyzed in this chapter. The properties of electrodes on which a reaction controlled by ion transport process takes place were taken into consideration. Also, it was analyzed how the electrode's shape and the voltage polarizing the electrode influence this electrode's metrological properties. The result of the analysis conducted is that in case of a reaction controlled by charge exchange process, such a voltammetric electrode functions like a converter type 0. Its metrological properties in the time domain are defined solely by sensitivity. However, if on the surface of the electrode there is a reaction controlled by ion transport process, the electrode will function like a converter type I. Its metrological properties in the time domain are defined by the sensitivity and time constant. Numeric simulations were conducted in order to determine the influence of the electrode's shape and the polarizing voltage on metrological properties of the electrode. The results show that both the sensitivity and the time constant of the electrode can be influenced by choice of an electrode's shape and the shape of the polarizing voltage.

Keywords: voltammetric electrodes, voltammetric measurements, DC voltammetry, AC voltammetry, metrologic properties, time constant

1. Introduction

Voltammetric measurements are one of the most frequently conducted measurements in order to determine ion concentration in water [1–7]. Their commonness is connected most of all to its simplicity and relatively high accuracy. There are many different types of voltammetric methods [7–15]. These methods differ from each other mainly in the voltage shape polarizing the voltammetric electrode, and in result, they also differ in accuracy of measurements conducted.

These methods are successfully applied in electrochemical measurements in stationary conditions when the marked ion concentration in the volume of analyzed solution is constant in time. At the same time, much more voltammetric measurements are conducted in situ, where the concentration of marked ions can change during the marking process [7, 10]. Some questions raise concerning the accuracy of conducted measurements, metrological properties of voltammetric electrodes, and methods of their improvement [12–15]. Hence, some work is undertaken in order to define metrological properties of voltammetric electrodes and the influence of the electrode's shape and the shape of polarizing voltage on these electrodes.

2. Metrological properties of voltammetric electrodes

Generally, metrological properties of voltammetric electrodes as measuring converters can be divided into static and dynamic ones.

Static properties are the characteristic of voltammetric electrodes which are in the steady state, i.e., in the state in which the concentration of marked ions does not change in the volume of the analyzed solution nor on the surface of the electrode. Dynamic properties are the characteristics of the electrode in the transient state, when these concentrations change while the voltammetric measurements are being conducted.

In order to simplify the analysis of the voltammetric electrode metrological properties following assumptions have been accepted:

- the input signal is the marked ion concentration $C_i^0(t)$ in the analyzed electrolyte volume,
- the output signal is the current $i_i(t)$ of the electrochemical reaction on the voltammetric electrode's surface,
- the time of charge exchange between the ions in the analyzed electrolyte and the voltammetric electrode equals 0.

Generally, the electrochemical reaction of oxidizing/reduction on the voltammetric electrode's surface is divided into several stages. The first stage is about delivery of depolarizer's ions from the volume of the electrolyte into the vicinity of the electrode's surface. The second stage of the electrochemical reaction is to transport, to or from, an electron or electrons through the depolarizer's ion. The third stage is to transport away the products from the reaction to the volume of analyzed solution. Stage four of the reaction is when the products of the electrochemical reaction can still react with other ions in the electrolyte after being transported away.

3. Electrochemical reaction controlled by a process of charge exchange on the surface of the voltammetric electrode

When on the surface of the voltammetric electrode an electrochemical reaction controlled by a process of charge exchange takes place, then the marked ion concentration in oxidizing/reduction

http://dx.doi.org/10.5772/67944

form on the electrode surface equals the concentration of the same ion form in the analyzed solution volume [8]:

$$C_{i,\,\mathrm{ox},0}(t) = C^0_{i,\,\mathrm{ox}}(t), \tag{1}$$

$$C_{i,\,\mathrm{red},0}(t) = C^0_{i,\,\mathrm{red}}(t). \tag{2}$$

The value of the output signal of a voltammetric electrode, which is the current on this electrode, is defined by the Butler-Volmer equation [8]:

$$i_i(t) = z_iFA[k_{i,\,\mathrm{ox}}(t)C_{i,\,\mathrm{ox},0}(t) - k_{i,\,\mathrm{red}}(t)C_{i,\,\mathrm{red},0}(t)] = i_{i,\,\mathrm{ox}}(t) - i_{i,\,\mathrm{red}}(t), \tag{3}$$

where the values of the reaction rate coefficients $k_{i,\,\mathrm{ox}}$ and $k_{i,\,\mathrm{red}}$ are defined by the following relations [8]:

$$k_{i,\,\mathrm{ox}}(t) = k^0 \exp\left\{-\frac{\alpha z_iF}{RT}\left[E_{\mathrm{pol}}(t) - E^0\right]\right\}, \tag{4}$$

$$k_{i,\,\mathrm{red}}(t) = k^0 \exp\left\{\frac{(1-\alpha)z_iF}{RT}\left[E_{\mathrm{pol}}(t) - E^0\right]\right\}. \tag{5}$$

Keeping in mind that an oxidizing/reduction reaction may take place on the surface of the voltammetric electrode, the relation (3) may be denoted as follows:

$$i_{i,\,\mathrm{ox}}(t) = z_iFAk_{i,\,\mathrm{ox}}(t)C_{i,\,\mathrm{ox},0}(t), \tag{6}$$

$$i_{i,\,\mathrm{red}}(t) = z_iFAk_{i,\,\mathrm{red}}(t)C_{i,\,\mathrm{red},0}(t). \tag{7}$$

It is immediately clear that the voltammetric electrode functions exactly like a converter type 0. Hence, its metrological properties are defined solely by the sensitivity coefficient denoted as follows:

$$S_i(t) = \frac{\Delta i_i(t)}{\Delta C_i^0(t)} = z_iFAk_i(t), \tag{8}$$

by taking into consideration relations (6) and (7) we get:

$$S_{i,\,\mathrm{ox}}(t) = z_iFAk_{i,\,\mathrm{ox}}(t), \tag{9}$$

$$S_{i,\,\mathrm{red}}(t) = z_iFAk_{i,\,\mathrm{red}}(t). \tag{10}$$

It results from the above analysis presented that metrological properties of a voltammetric electrode are described solely by sensitivity. This parameter is characteristic to an electrode both in the steady state and in the transient state. In such electrochemical reactions, the electrode does not present any delays or dynamic errors. Its sensitivity is determined by parameters defining marked ions, area of the electrode, electrochemical reaction rate on the surface of the electrode, and the voltage polarizing the electrode.

4. Electrochemical reaction controlled by a process of ion transport to the surface of the voltammetric electrode

In the case when an electrochemical reaction controlled by a process of ion transport takes place on the surface of the electrode, its flux to the surface is defined by this relation [8]:

$$N_i(t) = D_i \nabla C_i(t) + u_i z_i F C_i(t) \nabla U(t) + V_i(t) C_i(t). \tag{11}$$

Distribution of ion concentration $\nabla D_i = 0$ in the solution volume as a function of time t is defined by the flux divergence. Hence [8]:

$$\frac{\partial C_i(t)}{\partial t} = \nabla N_i(t). \tag{12}$$

Keeping in mind the relation (11) and assuming that $\nabla D_i = 0$ we get as a result:

$$\frac{\partial C_i(t)}{\partial t} = D_i \nabla^2 C_i(t) + z_i u_i F \nabla C_i(t) \nabla U(t) + z_i u_i F C_i(t) \nabla^2 U(t) + \nabla V_i(t) C_i(t) + V_i(t) \nabla C_i(t) \tag{13}$$

In real terms, voltammetric measurements are conducted with stationary electrodes in presence of excess of concentrated basic electrolyte, which allows to simplify the relation (13) to:

$$\frac{\partial C_i(t)}{\partial t} = D_i \nabla^2 C_i(t). \tag{14}$$

It is clear that in such a case the ion transport to or from the surface of a voltammetric electrode is determined solely by the diffusion.

4.1. Metrological properties for a general case

In general cases, without making any assumptions about reversibility or irreversibility of electrochemical reactions happening on the surface of the voltammetric electrode, one may present the relation (14) using the finite difference method

$$\frac{\partial C_{i,0}(t)}{\partial t} = D_i \left[\frac{C_i^0(t) - C_{i,0}(t)}{\delta_i^2(t)} \right], \tag{15}$$

Which, after transformation, leads to:

$$\left[\frac{\delta_i^2(t)}{D_i} \right] \frac{\partial C_{i,0}(t)}{\partial t} + C_{i,0}(t) = C_i^0(t) \tag{16}$$

Keeping in mind the relation (3) and transforming it we get:

http://dx.doi.org/10.5772/67944

$$C_{i,0}(t) = \frac{i_i(t)}{z_i F A k_i(t)}, \tag{17}$$

$$\frac{\partial C_{i,0}(t)}{\partial t} = \frac{1}{z_i F A k_i(t)} \left[\frac{\partial i_i(t)}{\partial t} - z_i F A C_{i,0}(t) \frac{\partial i_i(t)}{\partial t} \right]. \tag{18}$$

Substituting the relation (16) with relations (17) and (18) we get:

$$\left[\frac{\delta_i^2(t)}{D_i} \right] \frac{\partial i_i(t)}{\partial t} + \left[1 - \frac{\delta_i^2(t)}{D_i} \frac{1}{k_i(t)} \frac{\partial k_i(t)}{\partial t} \right] i_i(t) = z_i F A k_i(t) C_i^0(t). \tag{19}$$

Keeping in mind all the assumptions, we can transform the above relation to:

$$\frac{\left[\frac{\delta_i^2(t)}{D_i} \right]}{\left\{ 1 - \left[\frac{\delta_i^2(t)}{D_i} \right] \frac{1}{k_i(t)} \frac{\partial k_i(t)}{\partial t} \right\}} \frac{\partial i_i(t)}{\partial t} + i_i(t) = \frac{z_i F A k_i(t)}{\left\{ 1 - \left[\frac{\delta_i^2(t)}{D_i} \right] \frac{1}{k_i(t)} \frac{\partial k_i(t)}{\partial t} \right\}} C_i^0(t). \tag{20}$$

It results from the presented analysis that in this case the voltammetric electrode functions like converter type I. Its metrological properties are defined by sensitivity and the time constant. Static properties of the electrode are defined by sensitivity, and its dynamic properties are characterized by sensitivity and the time constant. Values of these parameters define the following relations:

$$S_{i,\text{ox}}(t) = \frac{z_i F A k_{i,\text{ox}}(t)}{\left\{ 1 - \left[\frac{\delta_{i,\text{ox}}^2(t)}{D_{i,\text{ox}}} \right] \frac{1}{k_{i,\text{ox}}(t)} \frac{\partial k_{i,\text{ox}}(t)}{\partial t} \right\}}, \tag{21}$$

$$N_{T,i,\text{ox}}(t) = \frac{\left[\frac{\delta_{i,\text{ox}}^2(t)}{D_{i,\text{ox}}} \right]}{\left\{ 1 - \left[\frac{\delta_{i,\text{ox}}^2(t)}{D_{i,\text{ox}}} \right] \frac{1}{k_{i,\text{ox}}(t)} \frac{\partial k_{i,\text{ox}}(t)}{\partial t} \right\}}, \tag{22}$$

$$S_{i,\text{red}}(t) = \frac{z_i F A k_{i,\text{red}}(t)}{\left\{ 1 - \left[\frac{\delta_{i,\text{red}}^2(t)}{D_{i,\text{red}}} \right] \frac{1}{k_{i,\text{red}}(t)} \frac{\partial k_{i,\text{red}}(t)}{\partial t} \right\}}, \tag{23}$$

$$N_{T,i,\text{red}}(t) = \frac{\left[\frac{\delta_{i,\text{red}}^2(t)}{D_{i,\text{red}}} \right]}{\left\{ 1 - \left[\frac{\delta_{i,\text{red}}^2(t)}{D_{i,\text{red}}} \right] \frac{1}{k_{i,\text{red}}(t)} \frac{\partial k_{i,\text{red}}(t)}{\partial t} \right\}}. \tag{24}$$

In the case of this type of electrochemical reactions, an electrode will present a dynamic error whose values are determined above all by the time constant and the nature of changes in the marked ion concentration. One can see in the presented relations that the parameters defining metrological properties of the voltammetric electrode are determined by the parameters defining marked ions, areas of the electrodes, a thickness of the diffusion layer, a rate of the electrochemical reaction happening on the surface of the electrode, and thereby voltage polarizing the electrode.

4.2. The influence of the voltammetric electrode's shape on its metrological properties in the time domain

Not only flat electrodes are being used in voltammetric measurements, but spherical and cylindrical as well. That is why we also analyzed the influence of the voltammetric electrode on its metrological properties.

4.2.1. Spherical voltammetric electrode

In the case of spherical voltammetric electrode, the relation of ion transport to/from the electrode's surface give in relation (14) appears as follows:

$$\frac{\partial C_{i,0}(t)}{\partial t} = D_i \left\{ \frac{1}{r^2}\frac{\partial}{\partial r}\left[r^2 \frac{\partial C_{i,0}(t)}{\partial r} \right] + \frac{1}{r^2 \sin\theta}\frac{\partial}{\partial \theta}\left[\sin\theta \frac{\partial C_{i,0}(t)}{\partial \theta} \right] + \frac{1}{r^2 \sin^2\theta}\frac{\partial^2 C_{i,0}(t)}{\partial \varphi^2} \right\} \tag{25}$$

Assuming that the marked ion concentration on the surface of the voltammetric electrode is not determined by θ and φ, so

$$\frac{\partial C_{i,0}(t)}{\partial \theta} = 0, \tag{26}$$

$$\frac{\partial^2 C_{i,0}(t)}{\partial \theta^2} = 0, \tag{27}$$

and

$$\frac{\partial C_{i,0}(t)}{\partial \varphi} = 0, \tag{28}$$

$$\frac{\partial^2 C_{i,0}(t)}{\partial \varphi^2} = 0, \tag{29}$$

relation of the ion transport given in relation (25) is simplified as follows:

$$\frac{\partial C_{i,0}(t)}{\partial t} = D_i \left\{ \frac{1}{r^2}\frac{\partial}{\partial r}\left[r^2 \frac{\partial C_{i,0}(t)}{\partial r} \right] \right\}, \tag{30}$$

and it results in:

$$\frac{\partial C_{i,0}(t)}{\partial t} = \frac{2D_i}{r}\frac{\partial C_{i,0}(t)}{\partial r} + D_i \frac{\partial^2 C_{i,0}(t)}{\partial r^2}. \tag{31}$$

We can present the relation using the finite difference method:

$$\frac{\partial C_{i,0}(t)}{\partial t} = \frac{2D_i}{r}\left[\frac{C_i^0(t) - C_{i,0}(t)}{\delta_i(t)}\right] + D_i\left[\frac{C_i^0(t) - C_{i,0}(t)}{\delta_i^2(t)}\right], \tag{32}$$

which after transformation leads to the relation:

$$\frac{\partial C_{i,0}(t)}{\partial t} + \left[\frac{2D_i\delta_i(t) + rD_i}{r\delta_i^2(t)}\right]C_{i,0}(t) = \left[\frac{2D_i\delta_i(t) + rD_i}{r\delta_i^2(t)}\right]C_i^0(t), \tag{33}$$

and it results in:

$$\frac{r\delta_i^2(t)}{2D_i\delta_i(t) + rD_i}\frac{\partial C_{i,0}(t)}{\partial t} + C_{i,0}(t) = C_i^0(t). \tag{34}$$

Substituting the relation (34) with relations (17) and (18) we get:

$$\frac{\partial i_i(t)}{\partial t} + \left[\frac{2D_i\delta_i(t) + rD_i}{r\delta_i^2(t)} - \frac{1}{k_i(t)}\frac{\partial k_i(t)}{\partial t}\right]i_i(t) = z_iFAk_i(t)\left[\frac{2D_i\delta_i(t) + rD_i}{r\delta_i^2(t)}\right]C_i^0(t), \tag{35}$$

which can be denoted as:

$$\frac{1}{\left[\frac{2D_i\delta_i(t)+rD_i}{r\delta_i^2(t)} - \frac{1}{k_i(t)}\frac{\partial k_i(t)}{\partial t}\right]}\frac{\partial i_i(t)}{\partial t} + i_i(t) = \frac{z_iFAk_i(t)\left[\frac{2D_i\delta_i(t)+rD_i}{r\delta_i^2(t)}\right]}{\left[\frac{2D_i\delta_i(t)+rD_i}{r\delta_i^2(t)} - \frac{1}{k_i(t)}\frac{\partial k_i(t)}{\partial t}\right]}C_i^0(t). \tag{36}$$

The above relation proves that the spherical voltammetric electrode functions as converter type I, both for the oxidizing reaction and for the reduction reaction. Its metrological properties are defined by sensitivity and the time constant. Static properties of the electrode are defined by sensitivity, and its dynamic properties are defined by sensitivity and the time constant. Values of these parameters are defined by the following relations:

$$S_{i,\mathrm{ox}}(t) = \frac{z_iFAk_{i,\mathrm{ox}}(t)\left[\frac{2D_{i,\mathrm{ox}}\delta_{i,\mathrm{ox}}(t)+rD_{i,\mathrm{ox}}}{r\delta_{i,\mathrm{ox}}^2(t)}\right]}{\left[\frac{2D_{i,\mathrm{ox}}\delta_{i,\mathrm{ox}}(t)+rD_{i,\mathrm{ox}}}{r\delta_{i,\mathrm{ox}}^2(t)} - \frac{1}{k_{i,\mathrm{ox}}(t)}\frac{\partial k_{i,\mathrm{ox}}(t)}{\partial t}\right]}, \tag{37}$$

$$N_{T,i,\mathrm{ox}}(t) = \frac{1}{\left[\frac{2D_{i,\mathrm{ox}}\delta_{i,\mathrm{ox}}(t)+rD_{i,\mathrm{ox}}}{r\delta_{i,\mathrm{ox}}^2(t)} - \frac{1}{k_{i,\mathrm{ox}}(t)}\frac{\partial k_{i,\mathrm{ox}}(t)}{\partial t}\right]}, \tag{38}$$

$$S_{i,\mathrm{red}}(t) = \frac{z_iFAk_{i,\mathrm{red}}(t)\left[\frac{2D_{i,\mathrm{red}}\delta_{i,\mathrm{red}}(t)+rD_{i,\mathrm{red}}}{r\delta_{i,\mathrm{red}}^2(t)}\right]}{\left[\frac{2D_{i,\mathrm{red}}\delta_{i,\mathrm{red}}(t)+rD_{i,\mathrm{red}}}{r\delta_{i,\mathrm{red}}^2(t)} - \frac{1}{k_{i,\mathrm{red}}(t)}\frac{\partial k_{i,\mathrm{red}}(t)}{\partial t}\right]}, \tag{39}$$

$$N_{T,i,\mathrm{red}}(t) = \frac{1}{\left[\frac{2D_{i,\mathrm{red}}\delta_{i,\mathrm{red}}(t)+rD_{i,\mathrm{red}}}{r\delta_{i,\mathrm{red}}^2(t)} - \frac{1}{k_{i,\mathrm{red}}(t)}\frac{\partial k_{i,\mathrm{red}}(t)}{\partial t}\right]}. \tag{40}$$

Presented relations show that a change of the shape of voltammetric electrode does not cause a change of the measuring converter's type. A spherical electrode functions as converter type I. This electrode, in this type of electrochemical reactions, will produce a dynamic error whose value is determined above all by the time constant and by the nature of changes of the marked ion concentration. The relations also show that the parameters defining metrological properties of this voltammetric electrode are determined by the parameters defining marked ions, an electrode radius, a thickness of the diffusion layer, a rate of electrochemical reaction taking place on the surface of the electrode, and thereby by the voltage polarizing the electrode.

4.2.2. Cylindrical voltammetric electrode

The relation of ion transport to/from the surface of the cylindrical voltammetric electrode shown by relation (14) is:

$$\frac{\partial C_{i,0}(t)}{\partial t} = D_i\left\{\frac{1}{\rho}\frac{\partial}{\partial \rho}\left[\rho\frac{\partial C_{i,0}(t)}{\partial \rho}\right] + \frac{1}{\rho^2}\frac{\partial^2 C_{i,0}(t)}{\partial \varphi^2} + \frac{\partial^2 C_{i,0}(t)}{\partial z^2}\right\}. \tag{41}$$

Assuming that the concentration of marked ions on the surface of the cylindrical electrode is determined by φ, i.e.,:

$$\frac{\partial C_{i,0}(t)}{\partial \varphi} = 0, \tag{42}$$

we get:

$$\frac{\partial C_{i,0}(t)}{\partial t} = \frac{D_i}{\rho}\frac{\partial C_{i,0}(t)}{\partial \rho} + D_i\frac{\partial^2 C_{i,0}(t)}{\partial \rho^2} + D_i\frac{\partial^2 C_{i,0}(t)}{\partial z^2}. \tag{43}$$

Defining this relation with finite difference method we get:

$$\frac{\partial C_{i,0}(t)}{\partial t} = \frac{D_i}{\rho}\left[\frac{C_i^0(t) - C_{i,0}(t)}{\delta_i(t)}\right] + D_i\left[\frac{C_i^0(t) - C_{i,0}(t)}{\delta_i^2(t)}\right] + D_i\left[\frac{C_i^0(t) - C_{i,0}(t)}{\delta_i^2(t)}\right], \tag{44}$$

which finally leads to:

$$\frac{\partial C_{i,0}(t)}{\partial t} = \left[\frac{D_i}{\rho\delta_i(t)} + \frac{2D_i}{\delta_i^2(t)}\right]C_{i,0}(t) = \left[\frac{D_i}{\rho\delta_i(t)} + \frac{2D_i}{\delta_i^2(t)}\right]C_i^0(t), \tag{45}$$

and it results in:

$$\frac{\rho\delta_i^2(t)}{2\rho D_i + \delta_i(t)D_i}\frac{\partial C_{i,0}(t)}{\partial t} + C_{i,0}(t) = C_i^0(t). \tag{46}$$

Substituting relation (46) with relations (17) and (18) we get:

$$\frac{\rho\delta_i^2(t)}{2\rho D_i + \delta_i(t)D_i}\frac{\partial i_i(t)}{\partial t} + \left[\frac{2\rho D_i + \delta_i(t)D_i}{\rho\delta_i^2(t)} - \frac{1}{k_i(t)}\frac{\partial k_i(t)}{\partial t}\right] i_i(t) = z_i F A k_i(t)\left[\frac{2\rho D_i + \delta_i(t)D_i}{\rho\delta_i^2(t)}\right] C_i^0(t). \tag{47}$$

Keeping in mind the assumptions, the above relation may be transformed as follows:

$$\frac{\left[\frac{\rho\delta_i^2(t)}{2\rho D_i+\delta_i(t)D_i}\right]}{\left[\frac{2\rho D_i+\delta_i(t)D_i}{\rho\delta_i^2(t)} - \frac{1}{k_i(t)}\frac{\partial k_i(t)}{\partial t}\right]}\frac{\partial i_i(t)}{\partial t} + i_i(t) = \frac{z_i F A k_i(t)\left[\frac{2\rho D_i+\delta_i(t)D_i}{\rho\delta_i^2(t)}\right]}{\left[\frac{2\rho D_i+\delta_i(t)D_i}{\rho\delta_i^2(t)} - \frac{1}{k_i(t)}\frac{\partial k_i(t)}{\partial t}\right]} C_i^0(t). \tag{48}$$

In the above relation, it is clear that a cylindrical voltammetric electrode functions as a measuring converter type I both for the oxidizing reaction and for the reduction reaction. Its metrological properties are defined by sensititvity and the time constant. Static properties of the electrode are defined by sensitivity, and its dynamic properties are characterized by sensitivity and the time constant. Values of the parameters are described as follows:

$$S_{i,\mathrm{ox}}(t) = \frac{z_i F A k_{i,\mathrm{ox}}(t)\left[\frac{2\rho D_{i,\mathrm{ox}}+\delta_{i,\mathrm{ox}}(t)D_{i,\mathrm{ox}}}{\rho\delta_{i,\mathrm{ox}}^2(t)}\right]}{\left[\frac{2\rho D_{i,\mathrm{ox}}+\delta_{i,\mathrm{ox}}(t)D_{i,\mathrm{ox}}}{\rho\delta_{i,\mathrm{ox}}^2(t)} - \frac{1}{k_{i,\mathrm{ox}}(t)}\frac{\partial k_{i,\mathrm{ox}}(t)}{\partial t}\right]}, \tag{49}$$

$$N_{T,i,\mathrm{ox}}(t) = \frac{\left[\frac{\rho\delta_{i,\mathrm{ox}}^2(t)}{2\rho D_{i,\mathrm{ox}}+\delta_{i,\mathrm{ox}}(t)D_{i,\mathrm{ox}}}\right]}{\left[\frac{2\rho D_{i,\mathrm{ox}}+\delta_{i,\mathrm{ox}}(t)D_{i,\mathrm{ox}}}{\rho\delta_{i,\mathrm{ox}}^2(t)} - \frac{1}{k_{i,\mathrm{ox}}(t)}\frac{\partial k_{i,\mathrm{ox}}(t)}{\partial t}\right]}, \tag{50}$$

$$S_{i,\mathrm{red}}(t) = \frac{z_i F A k_{i,\mathrm{red}}(t)\left[\frac{2\rho D_{i,\mathrm{red}}+\delta_{i,\mathrm{red}}(t)D_{i,\mathrm{red}}}{\rho\delta_{i,\mathrm{red}}^2(t)}\right]}{\left[\frac{2\rho D_{i,\mathrm{red}}+\delta_{i,\mathrm{red}}(t)D_{i,\mathrm{red}}}{\rho\delta_{i,\mathrm{red}}^2(t)} - \frac{1}{k_{i,\mathrm{red}}(t)}\frac{\partial k_{i,\mathrm{red}}(t)}{\partial t}\right]}, \tag{51}$$

$$N_{T,i,\mathrm{red}}(t) = \frac{\left[\frac{\rho\delta_{i,\mathrm{red}}^2(t)}{2\rho D_{i,\mathrm{red}}+\delta_{i,\mathrm{red}}(t)D_{i,\mathrm{red}}}\right]}{\left[\frac{2\rho D_{i,\mathrm{red}}+\delta_{i,\mathrm{red}}(t)D_{i,\mathrm{red}}}{\rho\delta_{i,\mathrm{red}}^2(t)} - \frac{1}{k_{i,\mathrm{red}}(t)}\frac{\partial k_{i,\mathrm{red}}(t)}{\partial t}\right]}. \tag{52}$$

In presented relations, one may notice that the change of the shape of a voltammetric electrode does not change its type as a measuring converter. A cylindrical electrode functions also as a converter type I. This electrode in case of this type of electrochemical reaction will create a dynamic error whose value is determined above all by the time constant and the nature of changes in marked ion concentration. In presented relations, one may notice that parameters defining metrological properties of the voltammetric electrode are determined by parameters of marked ions, an electrode radius, a thickness of the diffusion layer, a rate of the electrochemical reaction taking place on the surface of the electrode, and thereby by voltage polarizing the electrode.

4.3. The influence if polarizing voltage on metrological properties of a voltammetric electrode

In order to enforce a certain course of an electrochemical reaction on the surface of a voltammetric electrode, it should be polarized with a proper voltage. In measuring practice, there are different voltammetric methods applied. Most frequently used method is direct current voltammetry. In this method, an electrode used is polarized by voltage with a value changing linearly. The advantage of such a solution is the simplicity of the measuring system. At the same time, its disadvantage is relatively low accuracy. It is a solution connected with relatively great influence of capacitive current.

In order to eliminate the influence of the volume of the double layer on accuracy of voltammetric markings, there are different types of alternating currents voltammetry applied. In such cases, the voltage polarizing a voltammetric electrode has two components: variable and static. A static component is identical with one in the direct current voltammetry, and a variable component may be for example sinusoidal voltage, square wave voltage, or triangle wave voltage.

4.3.1. The influence of polarizing voltage with a value changing linearly

In the method of direct current voltammetry, a voltammetric electrode used for measuring is polarized with a voltage described as follows:

$$E_{\mathrm{pol}}(t) = E_0 \pm S_U t. \tag{53}$$

In such a case, coefficients of the rate of an electrochemical reaction which are defined by relations (4) and (5) will take on a form:

$$k_{i,\mathrm{ox}}(t) = k^0 \exp\left\{-\frac{\alpha z_i F}{RT}\left[E_0 + S_U t - E^0\right]\right\}, \tag{54}$$

$$k_{i,\mathrm{red}}(t) = k^0 \exp\left\{\frac{(1-\alpha) z_i F}{RT}\left[E_0 - S_U t - E^0\right]\right\}, \tag{55}$$

and their derivatives are, respectively:

$$\frac{\partial k_{i,\mathrm{ox}}(t)}{\partial t} = -\frac{\alpha z_i F}{RT} S_U k_{i,\mathrm{ox}}(t), \tag{56}$$

$$\frac{\partial k_{i,\mathrm{red}}(t)}{\partial t} = -\frac{(1-\alpha) z_i F}{RT} S_U k_{i,\mathrm{red}}(t). \tag{57}$$

Substituting relations (21) and (22) with relations (56) and (23) and relation (24) with relation (57) we get, respectively:

http://dx.doi.org/10.5772/67944

$$S_{i,\text{ox}}(t) = \frac{z_i F A k_{i,\text{ox}}(t)}{\left\{1 + \left[\frac{\delta^2_{i,\text{ox}}(t)}{D_{i,\text{ox}}}\right] \frac{\alpha z_i F}{RT} S_U\right\}}, \tag{58}$$

$$N_{T,i,\text{ox}}(t) = \frac{\left[\frac{\delta^2_{i,\text{ox}}(t)}{D_{i,\text{ox}}}\right]}{\left\{1 + \left[\frac{\delta^2_{i,\text{ox}}(t)}{D_{i,\text{ox}}}\right] \frac{\alpha z_i F}{RT} S_U\right\}}, \tag{59}$$

and:

$$S_{i,\text{red}}(t) = \frac{z_i F A k_{i,\text{red}}(t)}{\left\{1 + \left[\frac{\delta^2_{i,\text{red}}(t)}{D_{i,\text{red}}}\right] \frac{(1-\alpha) z_i F}{RT} S_U\right\}}, \tag{60}$$

$$N_{T,i,\text{red}}(t) = \frac{\left[\frac{\delta^2_{i,\text{red}}(t)}{D_{i,\text{red}}}\right]}{\left\{1 + \left[\frac{\delta^2_{i,\text{red}}(t)}{D_{i,\text{red}}}\right] \frac{(1-\alpha) z_i F}{RT} S_U\right\}}. \tag{61}$$

In relations above, one can see that a flat voltammetric electrode polarized with a voltage as in relation (53) functions as a converter type I. Parameters defining its metrological properties are determined by the rate of changes in polarizing voltage. Hence, their values can be alternated by an appropriate choice of the rate of changes in this voltage.

4.3.2. The influence of polarizing voltage with a sinusoidal variable component

In the method of alternating current sinusoidal voltammetry, the voltammetric electrode used for measuring is polarized with a voltage described this way:

$$E_{\text{pol}}(t) = E_0 \pm S_U t \pm U_m \sin \omega t. \tag{62}$$

In this case, coefficients of the rate of an electrochemical reaction described by the relations (4) and (5) are denoted:

$$k_{i,\text{ox}}(t) = k^0 \exp\left\{-\frac{\alpha z_i F}{RT}\left[E_0 + S_U t + U_m \sin \omega t - E^0\right]\right\}, \tag{63}$$

$$k_{i,\text{red}}(t) = k^0 \exp\left\{\frac{(1-\alpha) z_i F}{RT}\left[E_0 - S_U t - U_m \sin \omega t - E^0\right]\right\}, \tag{64}$$

and their derivatives are, respectively:

$$\frac{\partial k_{i,\text{ox}}(t)}{\partial t} = -\frac{\alpha z_i F}{RT}(S_U + \omega U_m \cos \omega t) k_{i,\text{ox}}(t), \tag{65}$$

$$\frac{\partial k_{i,\text{red}}(t)}{\partial t} = -\frac{(1-\alpha) z_i F}{RT}(S_U + \omega U_m \cos \omega t) k_{i,\text{red}}(t). \tag{66}$$

Because the variable component of polarizing voltage causes changes of ion concentration in the volume of the analyzed solution, also on the surface of the electrode, accordingly to the relation:

$$\frac{\partial[\Delta C_{i,0}(t)]}{\partial t} = \frac{\partial}{\partial t}[C_i^0(t)\sin\omega t] = \frac{\partial C_i^0(t)}{\partial t}\sin\omega t + \omega\cos\omega t C_i^0(t), \tag{67}$$

then the relation (14) denoted on the surface of the voltammetric electrode polarized by voltage with a static and sinusoidal variable component is:

$$\frac{\partial[C_{i,0}(t) + \Delta C_{i,0}(t)]}{\partial t} = \frac{\partial C_{i,0}(t)}{\partial t} + \frac{\partial[\Delta C_{i,0}(t)]}{\partial t} = D_i\nabla^2 C_i(t) + \frac{\partial C_i^0(t)}{\partial t}\sin\omega t + \omega\cos\omega t C_i^0(t). \tag{68}$$

Denoting this relation with the use of finite difference method we have:

$$\frac{\partial C_{i,0}(t)}{\partial t} = D_i\left[\frac{C_i^0(t) - C_{i,0}(t)}{\delta_i^2(t)}\right] + \frac{\partial C_i^0(t)}{\partial t}\sin\omega t + \omega\cos\omega t C_i^0(t), \tag{69}$$

which results in:

$$\left[\frac{\delta_i^2(t)}{D_i}\right]\frac{\partial C_{i,0}(t)}{\partial t} + C_{i,0}(t) = \left[\frac{\delta_i^2(t)}{D_i}\right]\left[\frac{D_i}{\delta_i^2(t)} + \omega\cos\omega t\right]\left\{C_i^0(t) + \frac{\frac{\partial C_i^0(t)}{\partial t}\sin\omega t}{\left[\frac{D_i}{\delta_i^2(t)} + \omega\cos\omega t\right]}\right\}. \tag{70}$$

Substituting the relation (70) with relations (17) and (18) we get:

$$\begin{gathered}\left[\frac{\delta_i^2(t)}{D_i}\right]\frac{\partial i_i(t)}{\partial t} + \left[1 - \frac{1}{k_i(t)}\frac{\partial k_i(t)}{\partial t}\frac{\delta_i^2(t)}{D_i}\right]i_i(t) = \left\{z_iFAk_i(t)\left[1 + \frac{\delta_i^2(t)}{D_i}\omega\cos\omega t\right]\right\} \\ \left\{C_i^0(t) + \frac{z_iFAk_i(t)\frac{\delta_i^2(t)}{D_i}\frac{\partial C_i^0(t)}{\partial t}\sin\omega t}{z_iFAk_i(t)\left[1 + \frac{\delta_i^2(t)}{D_i}\omega\cos\omega t\right]}\right\}.\end{gathered} \tag{71}$$

Keeping in mind the assumptions taken, we can transform the above relation into:

$$\begin{gathered}\frac{\left[\frac{\delta_i^2(t)}{D_i}\right]}{\left[1 - \frac{1}{k_i(t)}\frac{\partial k_i(t)}{\partial t}\frac{\delta_i^2(t)}{D_i}\right]}\frac{\partial i_i(t)}{\partial t} + i_i(t) = \frac{z_iFAk_i(t)\left[1 + \frac{\delta_i^2(t)}{D_i}\omega\cos\omega t\right]}{\left[1 - \frac{1}{k_i(t)}\frac{\partial k_i(t)}{\partial t}\frac{\delta_i^2(t)}{D_i}\right]} \\ \left\{C_i^0(t) + \frac{z_iFAk_i(t)\frac{\delta_i^2(t)}{D_i}\frac{\partial C_i^0(t)}{\partial t}\sin\omega t}{z_iFAk_i(t)\left[1 + \frac{\delta_i^2(t)}{D_i}\omega\cos\omega t\right]}\right\}.\end{gathered} \tag{72}$$

It is clear from the above relation that both for the oxidizing reaction and for the reduction reaction, a flat voltammetric electrode polarized by the voltage denoted as in relation (62) functions as a converter type I with properties defined as follows:

http://dx.doi.org/10.5772/67944

$$S_{i,\mathrm{ox}}(t) = \frac{z_i F A k_{i,\mathrm{ox}}(t)\left[1 + \frac{\delta_{i,\mathrm{ox}}^2(t)}{D_{i,\mathrm{ox}}}\omega\cos\omega t\right]}{\left[1 + \frac{\alpha z_i F}{RT}(S_U + \omega U_m \cos\omega t)\frac{\delta_{i,\mathrm{ox}}^2(t)}{D_{i,\mathrm{ox}}}\right]}, \tag{73}$$

$$N_{T,i,\mathrm{ox}}(t) = \frac{\frac{\delta_{i,\mathrm{ox}}^2(t)}{D_{i,\mathrm{ox}}}}{\left[1 + \frac{\alpha z_i F}{RT}(S_U + \omega U_m \cos\omega t)\frac{\delta_{i,\mathrm{ox}}^2(t)}{D_{i,\mathrm{ox}}}\right]}, \tag{74}$$

$$S_{i,\mathrm{red}}(t) = \frac{z_i F A k_{i,\mathrm{red}}(t)\left[1 + \frac{\delta_{i,\mathrm{red}}^2(t)}{D_{i,\mathrm{red}}}\omega\cos\omega t\right]}{\left[1 + \frac{(1-\alpha) z_i F}{RT}(S_U + \omega U_m \cos\omega t)\frac{\delta_{i,\mathrm{red}}^2(t)}{D_{i,\mathrm{red}}}\right]}, \tag{75}$$

$$N_{T,i,\mathrm{red}}(t) = \frac{\frac{\delta_{i,\mathrm{red}}^2(t)}{D_{i,\mathrm{red}}}}{\left[1 + \frac{(1-\alpha) z_i F}{RT}(S_U + \omega U_m \cos\omega t)\frac{\delta_{i,\mathrm{red}}^2(t)}{D_{i,\mathrm{red}}}\right]}. \tag{76}$$

It is clear that both the sensitivity of the electrode as well as its time constant are determined by the rate of changes of the static component of polarizing voltage and by the amplitude and sinusoidal frequency of the variable component. Hence, the parameters describing metrological properties of the electrode can be influenced by an appropriate choice of polarizing voltage parameters.

4.3.3. *The influence of the triangle waveform variable component*

In the method of triangular waveform AC voltammetry, a voltammetric electrode used for measuring is polarized by voltage denoted as follows:

$$E_{\mathrm{pol}}(t) = E_0 \pm S_U t \pm U_t(t), \tag{77}$$

with a variable component of the polarizing voltage which can be denoted as:

$$U_t(t) = \begin{cases} \dfrac{2At}{\pi} & -\dfrac{T}{2} \le t \le \dfrac{T}{2} \\ \dfrac{2A(\pi - t)}{\pi} & \dfrac{T}{2} \le t \le T \end{cases}. \tag{78}$$

Triangular bipolar waveform component $U_t(t)$ can be denoted with an expansion in the Fourier series:

$$U_t(t) = \frac{8A}{\pi^2}\sum_{n=0}^{\infty}(-1)^n \frac{\sin[(2n+1)\omega t]}{(2n+1)^2} = \frac{8A}{\pi^2}\left(\sin\omega t - \frac{1}{9}\sin 3\omega t + \frac{1}{25}\sin 5\omega t - \dots\right). \tag{79}$$

In this case coefficients of the rate of electrochemical reactions denoted by the relations (4) and (5) are as follows:

$$k_{i,\,\mathrm{ox}}(t) = k^0 \exp\left\{-\frac{\alpha z_i F}{RT}\left[E_0 + S_U t + \frac{8A}{\pi^2}\left(\sin\omega t - \frac{1}{9}\sin 3\omega t + \frac{1}{25}\sin 5\omega t - \ldots\right) - E^0\right]\right\}, \tag{80}$$

$$k_{i,\,\mathrm{red}}(t) = k^0 \exp\left\{\frac{(1-\alpha) z_i F}{RT}\left[E_0 - S_U t - \frac{8A}{\pi^2}\left(\sin\omega t - \frac{1}{9}\sin 3\omega t + \frac{1}{25}\sin 5\omega t - \ldots\right) - E^0\right]\right\}, \tag{81}$$

and their derivatives are, respectively:

$$\frac{\partial k_{i,\,\mathrm{ox}}(t)}{\partial t} = -\frac{\alpha z_i F}{RT}\left[S_U + \frac{8\omega A}{\pi^2}\left(\cos\omega t - \frac{1}{3}\cos 3\omega t + \frac{1}{5}\cos 5\omega t - \ldots\right)\right] k_{i,\,\mathrm{ox}}(t), \tag{82}$$

$$\frac{\partial k_{i,\,\mathrm{red}}(t)}{\partial t} = -\frac{(1-\alpha) z_i F}{RT}\left[S_U + \frac{8\omega A}{\pi^2}\left(\cos\omega t - \frac{1}{3}\cos 3\omega t + \frac{1}{5}\cos 5\omega t - \ldots\right)\right] k_{i,\,\mathrm{red}}(t). \tag{83}$$

Because the variable component of the polarizing voltage causes changes in the ion concentration of the analyzed solution, also on the surface of the electrode, accordingly to the relation:

$$\frac{\partial[\Delta C_{i,0}(t)]}{\partial t} = \frac{\partial}{\partial t}\left[\frac{8A}{\pi^2} C_i^0(t) \sum_{n=0}^{\infty} (-1)^n \frac{\sin[(2n+1)\omega t]}{(2n+1)^2}\right], \tag{84}$$

which leads to:

$$\frac{\partial[\Delta C_{i,0}(t)]}{\partial t} = \frac{8A}{\pi^2}\frac{\partial C_i^0(t)}{\partial t} \sum_{n=0}^{\infty} (-1)^n \frac{\sin[(2n+1)\omega t]}{(2n+1)^2} + \frac{8A}{\pi^2} C_i^0(t) \sum_{n=0}^{\infty} (-1)^n \frac{\omega\cos[(2n+1)\omega t]}{(2n+1)}, \tag{85}$$

then the relation (14) denoted for the surface of the voltammetric electrode polarized by the voltage with a static component and a triangular waveform variable component is:

$$\begin{aligned}\frac{\partial[C_{i,0}(t) + \Delta C_{i,0}(t)]}{\partial t} &= D_i \nabla^2 C_i(t) \\ &+ \frac{8A}{\pi^2}\left[\left(\sin\omega t - \frac{1}{9}\sin 3\omega t + \frac{1}{25}\sin 5\omega t - \ldots\right)\frac{\partial C_i^0(t)}{\partial t}\right. \\ &\left. + \omega C_i^0(t)\left(\cos\omega t - \frac{1}{3}\cos 3\omega t + \frac{1}{5}\cos 5\omega t - \ldots\right)\right].\end{aligned} \tag{86}$$

Denoting this relation by using finite difference method, we get:

$$\begin{aligned}\frac{\partial C_{i,0}(t)}{\partial t} &= D_i\left[\frac{C_i^0(t) - C_{i,0}(t)}{\delta_i^2(t)}\right] \\ &+ \frac{8A}{\pi^2}\left[\left(\sin\omega t - \frac{1}{9}\sin 3\omega t + \frac{1}{25}\sin 5\omega t - \ldots\right)\frac{\partial C_i^0(t)}{\partial t}\right. \\ &\left. + \omega C_i^0(t)\left(\cos\omega t - \frac{1}{3}\cos 3\omega t + \frac{1}{5}\cos 5\omega t - \ldots\right)\right],\end{aligned} \tag{87}$$

which results in:

$$\left[\frac{\delta_i^2(t)}{D_i}\right]\frac{\partial C_{i,0}(t)}{\partial t}+C_{i,0}(t)=\left[\frac{\delta_i^2(t)}{D_i}\right]$$
$$\left[\frac{D_i}{\delta_i^2(t)}+\frac{8\omega A}{\pi^2}\left(\cos\omega t-\frac{1}{3}\cos 3\omega t+\frac{1}{5}\cos 5\omega t-\dots\right)\right]$$
$$\left\{C_i^0(t)+\frac{\frac{8A}{\pi^2}\frac{\partial C_i^0(t)}{\partial t}\left(\sin\omega t-\frac{1}{9}\sin 3\omega t+\frac{1}{25}\sin 5\omega t-\dots\right)}{\left[\frac{D_i}{\delta_i^2(t)}+\frac{8\omega A}{\pi^2}\left(\cos\omega t-\frac{1}{3}\cos 3\omega t+\frac{1}{5}\cos 5\omega t-\dots\right)\right]}\right\}. \tag{88}$$

Substituting the relation (88) with relations (17) and (18), we get the relation:

$$\left[\frac{\delta_i^2(t)}{D_i}\right]\frac{\partial i_i(t)}{\partial t}+\left[1-\frac{1}{k_i(t)}\frac{\partial k_i(t)}{\partial t}\frac{\delta_i^2(t)}{D_i}\right]i_i(t)=$$
$$\left\{z_iFAk_i(t)\left[1+\frac{\delta_i^2(t)}{D_i}\frac{8\omega A}{\pi^2}\left(\cos\omega t-\frac{1}{3}\cos 3\omega t+\frac{1}{5}\cos 5\omega t-\dots\right)\right]\right\}$$
$$\left\{C_i^0(t)+\frac{z_iFAk_i(t)\frac{\delta_i^2(t)}{D_i}\frac{8A}{\pi^2}\frac{\partial C_i^0(t)}{\partial t}\left(\sin\omega t-\frac{1}{9}\sin 3\omega t+\frac{1}{25}\sin 5\omega t-\dots\right)}{z_iFAk_i(t)\left[1+\frac{\delta_i^2(t)}{D_i}\frac{8\omega A}{\pi^2}\left(\cos\omega t-\frac{1}{3}\cos 3\omega t+\frac{1}{5}\cos 5\omega t-\dots\right)\right]}\right\}. \tag{89}$$

Keeping in mind the assuptions taken, the above relation may be transformed into:

$$\frac{\left[\frac{\delta_i^2(t)}{D_i}\right]}{\left\{1+\frac{\delta_i^2(t)}{D_i}\frac{\alpha z_iF}{RT}\left[S_U+\frac{8\omega A}{\pi^2}\left(\cos\omega t-\frac{1}{3}\cos 3\omega t+\frac{1}{5}\cos 5\omega t-\dots\right)\right]\right\}}\frac{\partial i_i(t)}{\partial t}+i_i(t)=$$
$$\frac{z_iFAk_i(t)\left[1+\frac{\delta_i^2(t)}{D_i}\frac{8\omega A}{\pi^2}(\cos\omega t+\cos 3\omega t+\cos 5\omega t+\dots)\right]}{\left\{1+\frac{\delta_i^2(t)}{D_i}\frac{\alpha z_iF}{RT}\left[S_U+\frac{8\omega A}{\pi^2}\left(\cos\omega t-\frac{1}{3}\cos 3\omega t+\frac{1}{5}\cos 5\omega t-\dots\right)\right]\right\}}$$
$$\left\{C_i^0(t)+\frac{z_iFAk_i(t)\frac{\delta_i^2(t)}{D_i}\frac{8A}{\pi^2}\frac{\partial C_i^0(t)}{\partial t}\left(\sin\omega t-\frac{1}{9}\sin 3\omega t+\frac{1}{25}\sin 5\omega t-\dots\right)}{z_iFAk_i(t)\left[1+\frac{\delta_i^2(t)}{D_i}\frac{8\omega A}{\pi^2}\left(\cos\omega t-\frac{1}{3}\cos 3\omega t+\frac{1}{5}\cos 5\omega t-\dots\right)\right]}\right\}. \tag{90}$$

It is clear from the above relation that both for the oxidizing reaction and for the reduction reaction, a flat voltammetric electrode polarized by voltage as in relation (77) functions as a converter type I with properties defined as follows:

$$S_{i,\text{ox}}(t) = \frac{z_i F A k_{i,\text{ox}}(t)\left[1 + \frac{\delta_{i,\text{ox}}^2(t)}{D_{i,\text{ox}}}\frac{8\omega A}{\pi^2}\left(\cos\omega t + \cos 3\omega t + \cos 5\omega t + \ldots\right)\right]}{\left\{1 + \frac{\delta_{i,\text{ox}}^2(t)}{D_{i,\text{ox}}}\frac{\alpha z_i F}{RT}\left[S_U + \frac{8\omega A}{\pi^2}\left(\cos\omega t - \frac{1}{3}\cos 3\omega t + \frac{1}{5}\cos 5\omega t - \ldots\right)\right]\right\}}, \quad (91)$$

$$N_{T,i,\text{ox}}(t) = \frac{\left[\frac{\delta_{i,\text{ox}}^2(t)}{D_{i,\text{ox}}}\right]}{\left\{1 + \frac{\delta_{i,\text{ox}}^2(t)}{D_{i,\text{ox}}}\frac{\alpha z_i F}{RT}\left[S_U + \frac{8\omega A}{\pi^2}\left(\cos\omega t - \frac{1}{3}\cos 3\omega t + \frac{1}{5}\cos 5\omega t - \ldots\right)\right]\right\}}, \quad (92)$$

$$S_{i,\text{red}}(t) = \frac{z_i F A k_{i,\text{red}}(t)\left[1 + \frac{\delta_{i,\text{red}}^2(t)}{D_{i,\text{red}}}\frac{8\omega A}{\pi^2}\left(\cos\omega t + \cos 3\omega t + \cos 5\omega t + \ldots\right)\right]}{\left\{1 + \frac{\delta_{i,\text{red}}^2(t)}{D_{i,\text{red}}}\frac{(1-\alpha) z_i F}{RT}\left[S_U + \frac{8\omega A}{\pi^2}\left(\cos\omega t - \frac{1}{3}\cos 3\omega t + \frac{1}{5}\cos 5\omega t - \ldots\right)\right]\right\}}, \quad (93)$$

$$N_{T,i,\text{red}}(t) = \frac{\left[\frac{\delta_{i,\text{red}}^2(t)}{D_{i,\text{red}}}\right]}{\left\{1 + \frac{\delta_{i,\text{red}}^2(t)}{D_{i,\text{red}}}\frac{(1-\alpha) z_i F}{RT}\left[S_U + \frac{8\omega A}{\pi^2}\left(\cos\omega t - \frac{1}{3}\cos 3\omega t + \frac{1}{5}\cos 5\omega t - \ldots\right)\right]\right\}}. \quad (94)$$

It is clear that the sensitivity of an electrode and its time constant are determined by the rate of changes of the static component in the polarizing voltage and by the amplitude and frequency of the fundamental component and individual harmonic components of the triangle waveform polarizing voltage.

Hence, the parameters describing metrological properties of an electrode can be influenced by an appropriate choice of polarizing voltage parameters.

5. Numerical simulations and discussion

Numeric simulations were conducted in order to determine how the electrode's shape and the shape of the voltage influencing the voltammetric electrode metrological properties.

The influence of the voltammetric electrode's shape on its metrological properties was analyzed assuming that the electrode is flat, spherical, and cylindrical and polarized only by linearly increasing voltage.

Also, the influence of the shape of the voltage polarizing the electrode on its metrological properties was analyzed assuming that the electrode is flat and polarized by the linearly increasing voltage, linearly increasing voltage with sinusoidal variable component and linearly increasing voltage with a triangular waveform variable component. It was assumed in the simulations that there is an oxidizing reaction of marked ions on the surface of the electrode and the values defining the marked ions, voltammetric electrode's shapes and polarizing voltages are: $z_i = 1$, $\alpha = 0.5$, $E_0 = 0.25\ldots0.75$ V; $D_i = 10^{-5}\ldots10^{-7}$ cm/s, $A = 1$ mm^2, $r = 0.1\ldots1$ mm, $\rho = 0.1\ldots1$ mm, $S_U = 5\ldots100$ mV/s, $U_m = 5\ldots100$ mV, $f = 1\ldots100$ Hz.

Numerical simulations prove that the shape of the electrode and its geometrical dimensions influence its sensitivity. When the electrode functions like a converter type 0, the geometrical values influence the sensitivity solely through the surface area of the electrode. However, if the electrode functions like a converter type I, the electrode's radius influences greatly the sensitivity. The sensitivity of the electrode is the highest when the electrode functions like a converter type 0, which is shown in **Figure 1**.

The highest sensitivity is the characteristic of a cylindrical electrode when there are determined: rate of changes in the voltage polarizing an electrode, a rate of oxidizing reaction, an ion diffusion coefficient, and equal geometrical dimensions. It was also proved that spherical and cylindrical electrodes sensitivity is determined by their radius. Reducing geometrical dimensions of a flat voltammetric electrode leads to reduction of its sensitivity. And reducing the radius of spherical and cylindrical electrode leads to an increase of their sensitivity.

Numeric simulation results show that the shape of the electrode influences its time constant. The lowest time constant is a characteristic of a cylindrical electrode when there are determined: rate of changes in the voltage polarizing an electrode, an ion diffusion coefficient, and equal geometrical dimensions of electrodes, which is shown in **Figure 2**.

The results of calculations show that the shape of the voltage polarizing a flat voltammetric electrode influences its metrological properties. It has been proved that a flat electrode polarized by voltage with a triangular waveform variable component has the highest sensitivity when there are determined: an ion diffusion coefficient, steady rate of the reaction, equal parameters of the polarizing voltage, which is shown in **Figure 3**.

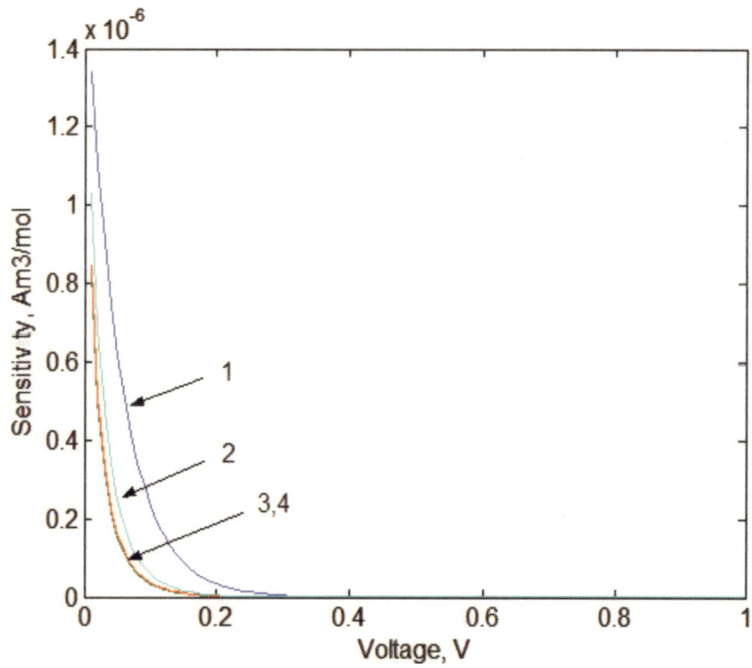

Figure 1. Influence of the shape on the sensitivity of an electrode; $E^0 = 0.50$ V; $D_i = 10^{-6}\text{cm}^2/\text{s}$, $k_0 = 10^{-8}$ cm/s, $A = 1\ \text{mm}^2$, $r = 0.5$ mm, $\rho = 0.5$ mm, $S_U = 0.025$ mV/s; The type of an electrode: (1) converter type 0; converter type I, (2) cylindrical, (3) spherical, (4) flat.

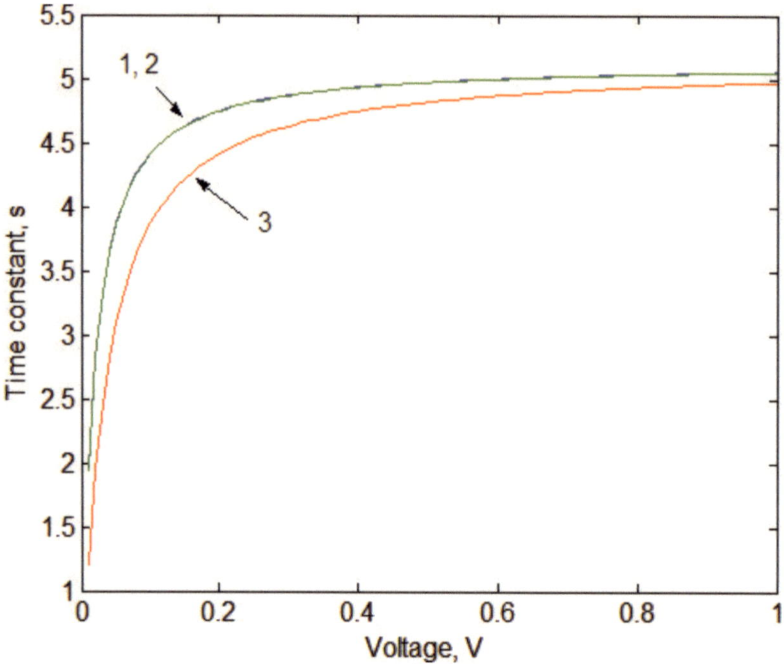

Figure 2. Influence of the shape on the time constant of an electrode; $D_i = 10^{-7}\text{cm}^2/\text{s}$, $r = 0.5$ mm, $\rho = 0.5$ mm, $S_U = 0.010$ mV/s; The type of an electrode: (1) flat, (2) spherical, (3) cylindrical.

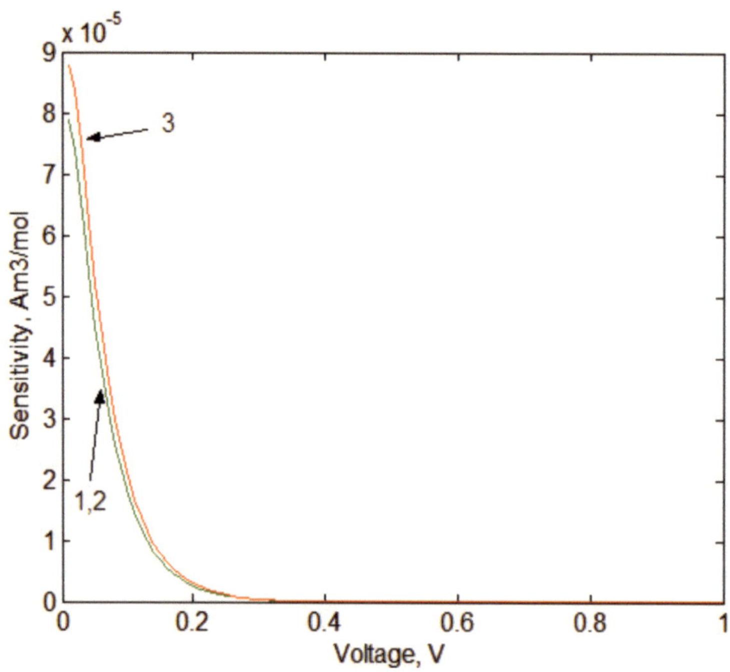

Figure 3. Influence of the shape of polarization voltage on the sensitivity of an electrode; $D_i = 10^{-7}\text{cm}^2/\text{s}$, $S_U = 0.010$ mV/s; $U_m = 5$ mV, $f = 1$ Hz; The type of polarization voltage: (1) direct voltage, (2) direct voltage with a sinusoidal component, (3) direct voltage with a triangular component.

The sensitivity of an electrode polarized with such a type of voltage is determined by the number of harmonics of the electrode's triangular waveform variable component. Including in calculations further, harmonics of this component increases the sensitivity of the electrode. The results of such calculations show that the sensitivity of a flat voltammetric electrode is also determined by the frequency of a polarizing voltage variable component. The increase of a frequency of sinusoidal and triangular waveform component voltage leads to the increase of a voltammetric electrode's sensitivity. The lower the rate of change of polarizing voltage static component the greater the increase of sensitivity. The amplitude of the polarizing voltage variable component influences the sensitivity of the electrode as well. The increase of this component's amplitude leads to the reduction of the electrode's sensitivity.

It was also proved that the time constant of a flat voltammetric electrode is determined by the shape of polarizing voltage. Results of these calculations revealed that the electrode polarized by voltage only with a static component has the highest time constant. The increase of the rate of changes in this voltage leads to the decrease of time constant of the electrode. A flat voltammetric electrode has the lowest time constant for voltage with a sinusoidal variable component when there are determined: rate of changes of the polarizing voltage static component and amplitude and frequency of the variable component, which is shown in **Figure 4**.

The increase of the variable component amplitude leads to the decrease of the voltammetric electrode time constant. Also an increase of a polarizing voltage variable component frequency leads to the decrease of the electrode's time constant. In the case when a flat electrode is polarized by voltage with a triangular waveform variable component including this electrode's individual

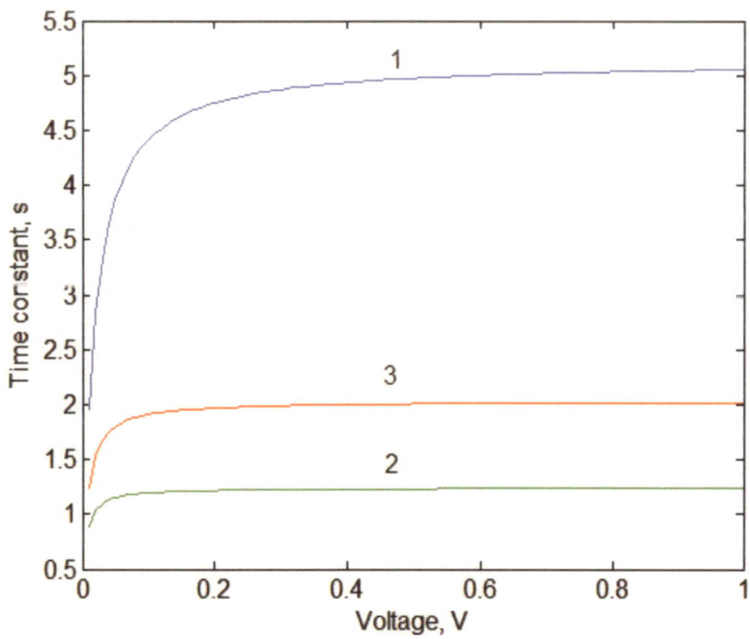

Figure 4. Influence of the shape on the time constant of a flat electrode; $D_i = 10^{-7} cm^2/s$, $r = 0.5$ mm, $\rho = 0.5$ mm, $S_U = 0.010$ mV/s; $U_m = 5$ mV, $f = 1$ Hz. The type of polarization voltage: (1) direct voltage, (2) direct voltage with a sinusoidal component, (3) direct voltage with a triangular component.

harmonics leads to the decrease of the time constant. The electrode's time constant also decreases when the rate of gain of the electrode polarizing voltage static component increases.

6. Conclusions

Analyses conducted show that depending on the concentration of marked ion solution, the voltammetric electrode may function like a converter type 0 or type I. In the first case, its metrological properties in the time domain are determined solely by sensitivity, in the latter case by sensitivity and the time constant. Values of both parameters are determined by the type of marked ions in the analyzed solution, by the shape of the electrode and the shape of the polarizing voltage. Results of simulations conducted show that in the case of marking low ion concentrations, the cylindrical electrode has the best metrological properties. It is characterized by the highest sensitivity and the lowest time constant. It was also proved that the most beneficial metrological properties are the ones of a voltammetric electrode polarized by voltage with a static component and a sinusoidal variable component. The choice of the amplitude and the frequency of polarized voltage variable component may influence both sensitivity and the time constant of a voltammetric electrode used in markings. As a result of such activities metrological properties of a used measuring electrode may be shaped in an optional way. The ion concentration may be measured with a high sensitivity or low time constant. It will allow to match closer metrological properties of a voltammetric electrode to the expected nature of changes in marked ion concentration and as a consequence it leads to the decrease of static and dynamic errors of electrochemical markings conducted.

Major symbols

A	Surface area of electrode
$C_i^0(t)$	Bulk concentration of ions i at time t
$C_{i,0}(t)$	Surface concentration of ions i at the electrode surface at time t
$\frac{\partial C_i(x)}{\partial t}$	Gradient of concentration of ions i at time t at voltammetric electrode
$\nabla C_i(x)$	Concentration gradient of ions i at distance x
$\nabla C_i(x)$	Concentration gradient of ions i at distance x
$C_{i,\mathrm{ox},0}(t)$	Concentration of oxidation ions i at the electrode surface at time t
$C_{i,\mathrm{red},0}(t)$	Concentration of reduction ions i at the electrode surface at time t
D_i	Diffusion coefficient of ions i
E^0	Standard potential of an electrode
E_0	Initial potential
F	The Faraday constant
$i(t)$	Current
$i_{\mathrm{ox}}(t)$	Anodic oxidation current
$i_{\mathrm{red}}(t)$	Cathodic reduction current

$k_{i,\mathrm{ox}}(t)$	Heterogeneous rate constant for oxidation
$k_{i,\mathrm{red}}(t)$	Heterogeneous rate constant for reduction
k^0	Rate constant for a heterogeneous reaction
$\nabla N_i(t)$	Flux gradient of ions *i*
u_i	Mobility of ions *i*
z_i	Charge on ions *i* in signed units of electronic charge
S_U	Linear potential scan rate
t	Time
T	Absolute temperature
U_m	Amplitude of a sinusoidal component
$U_t(t)$	Amplitude of a triangle component
∇U	Potential gradient
V_i	Velocity of ions *i*
α	Transfer coefficient
ω	Angular frequency of rotation
$\delta_i(t)$	Diffusion layer thickness for ions *i* at the electrode surface at time *t*
$\delta_{i,\mathrm{ox}}(t)$	Diffusion layer thickness for oxidation ions *i* at the electrode surface at time *t*
$\delta_{i,\mathrm{red}}(t)$	Diffusion layer thickness for reduction ions *i* at the electrode surface at time *t*

Author details

Krzysztof Suchocki

Address all correspondence to: krzsucho@pg.gda.pl

Department of Biomedical Engineering, Technical University of Gdansk, Gdansk, Poland

References

[1] Pierre MB, Jacques B. Polarography, voltammetry and tensammetry: tools for day-to-day analysis in the industrial laboratory. Analyst. 1989; **114**(12):1531–1544. DOI : 10.1039/AN9891401531

[2] Robert IM, Alan MB, Terence JC, Robert WC, Roger WK, Michael GNO, Bruce RC On-line and off-line voltammetric methods for the determination of nickel in zinc plant electrolyte. Analyst. 1994; **119**(5):1057–1061. DOI: 10.1039/AN994190105

[3] Alan MB. Past, present and future contributions of microelectrodes to analytical studies employing voltammetric detection. A review: Analyst. 1994, **119**(11):1R–21R. DOI: 10.1039/AN994190001R

[4] Williams G, D'Silva C. Field-based heavy metal analyser for the simultaneous determination of multiple cations on-site. Analyst. 1994; **119**(11):2337–2341. DOI: 10.1039/AN9941902337

[5] Damien WMA. Tutorial review. Voltammetric determination of trace metals and organics after accumulation at modified electrodes. Analyst. 1994; **119**(9):1953–1966. DOI: 10.1039/AN9941901953

[6] Roy MH. Understanding Our Environment, an Introduction to Environmental, Chemistry and Pollution. 3rd ed. Royal Society of Chemistry; 1999. p. 485. ISBN 0-85404-584-8. DOI:10.1039/9781847552235-00071.

[7] Allen JB, Larry RF. Electrochemical Methods: Fundamentals and Applications. 2nd ed. New York: John Wiley & Sons Inc.; 2001. p. 864. ISBN: 978-0-471-04372-0

[8] Patrycja C, Wojciech W. Sensor arrays for liquid sensing – electronic tongue systems. Analyst. 2007; **132**(10):963–978

[9] Wassana Y, Kitiya H, Cynthia LW, Daiwon C, Thanapon S, Mychailo BT, Marvin GW, Glen EF, Shane RA, Charles T. Direct detection of Pb in urine and Cd, Pb, Cu, and Ag in natural waters using electrochemical sensors immobilized with DMSA functionalized magnetic nanoparticles. Analyst. 2008; **133**(3):348–355. DOI: 10.1039/B711199A

[10] Niina JR, Brian HH, William RH. Electrochemical biosensors. Chemical Society Reviews. 2010; **39**(5):1747–1763. DOI: 10.1039/B714449K

[11] Jonathan PM, Rashid OK, Craig EB. New directions in screen printed electroanalytical sensors: an overview of recent developments. Analyst. 2011; **136**(6):1067–1076. DOI: 10.1039/C0AN00894J

[12] Rongsheng C, Yong L Kaifu H, Paul KC. Microelectrode arrays based on carbon nanomaterials: emerging electrochemical sensors for biological and environmental applications. Royal Society of Chemistry Advances. 2013; **3**(41):18698–18715. DOI: 10.1039/C3RA43033B

[13] Jonathan PM, Edward PR, Craig EB. Screen-printed back-to-back electroanalytical sensors. Analyst. 2014;**139**(21):5339–5349. DOI: 10.1039/C4AN01501K

[14] de la MGM, Rosario MPM, Lorenzo CB, Eduardo PG. Applicability of the bismuth bulk rotating disk electrode for heavy metal monitoring in undisturbed environmental and biological samples: determination of Zn(II) in rainwater, tap water and urine. Analytical Methods. 2014; **6**(21):8668–8674. DOI: 10.1039/C4AY01626B

[15] Abdel-Nasser K, Nadeem B, Muhammad S. Graphite pencil electrodes as electrochemical sensors for environmental analysis: a review of features, developments, and applications. Royal Society of Chemistry Adavances. 2016; **6**(94):91325–91340. DOI: 10.1039/C6RA17466C

Section 2

Voltammetric Analysis

Chapter 3

Direct Electron Transfer of Human Hemoglobin Molecules on Glass/Tin-Doped Indium Oxide

Flavio Dolores Martínez-Mancera and
José Luis Hernández-López

Additional information is available at the end of the chapter

http://dx.doi.org/10.5772/67806

Abstract

Interfacial electron transfer kinetics of the *haem* (Fe^{III}/Fe^{II}) group in human hemoglobin molecules were investigated on glass/tin-doped indium oxide electrodes. Factors such as surface roughness, crystallinity, hydrophilicity and partial polarization of the working electrode played an important role to provide a more compatible microenvironment for protein adsorption. Results suggested that direct electron transfer from electrode to *haem* (Fe^{III})-H_2O intermediate is coupled to proton at near physiological pH ($I = 0.035$, $pH = 7.2$).

Keywords: cyclic voltammetry, direct-electron-transfer, human hemoglobin, tin-doped indium oxide electrode, surface electron transfer rate constant

1. Introduction

Haem-containing proteins such as hemoglobin (Hb), also spelled haemoglobin, are macromolecules that consist in an assembly of four globular polypeptide chains, tightly associated with a nonprotein *haem* group by means a complex arrangement folding pattern (α-helix). The *haem* group consists of an iron atom chelated to a porphyrin ring (*cf.* **Figure 1**), which allow to carry the oxygen in the red blood cells to whole body of all vertebrates as well as some invertebrates. Although the iron atom can take any of its oxidation states (Fe^{II} or Fe^{III}), the ferrihemoglobin (methemoglobin, metHb) (Fe^{III}) cannot bind oxygen [1]. In adult humans, the most common Hb type is a tetramer well-known as Hb A, consisting of two α and two β subunits noncovalently bound ($\alpha_2\beta_2$). These subunits are structurally similar to themselves and about the same molecular size. The total molecular weight of the Hb A is *ca.* 64 kDa.

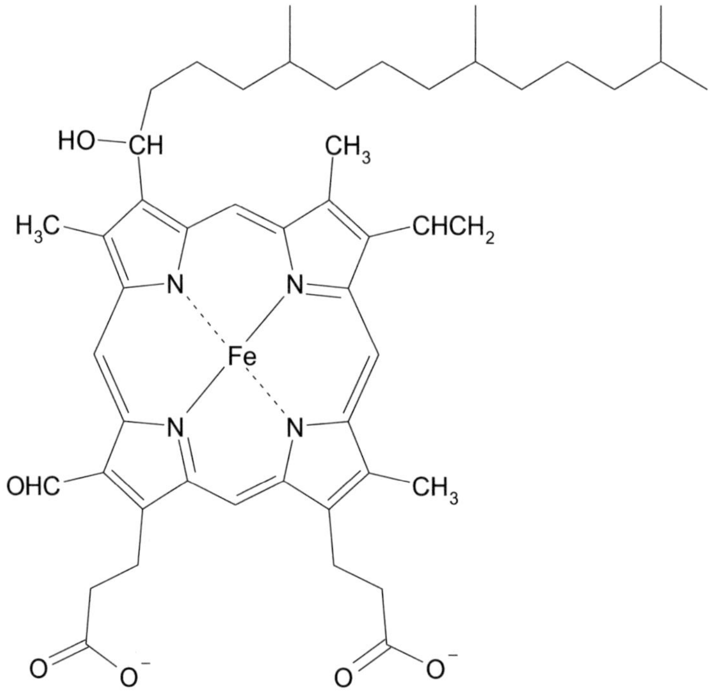

Figure 1. The structure of *haem* a.

The four polypeptide chains are bound to each other by salt bridges, hydrogen bonds, and hydrophobic interaction. While Hb does not function physiologically as an electron transfer carrier, it does undergo oxidation and reduction at the *haem* group in certain cases *in vivo* [2]. Therefore, the focused research on its electron transfer process might lead to a more profound understanding of electron flow in biological systems.

Throughout almost half of the century, there has been shown that the direct electrochemistry of *haem* proteins on bare electrodes is fairly difficult [3]. Arrival to this conclusion may be caused by several factors that were overcame to progress, among them: (a) the extended three-dimensional protein conformation due to strong interaction between the protein and the substrate or the lack of an effective microenvironment for adsorption; (b) the inaccessibility of electron communication between the electroactive center of the protein and the electrode due to misalignment of the redox center of the protein; (c) the adsorption of denatured protein onto electrodes, resulting in a loss of bioactivity; and (d) the unsymmetrical distribution of surface charges on protein molecules. According to point (c), a general problem commonly found with used metal electrodes, such as Au, Ag, Pt, and Hg, is that all of them lead to denaturation and irreversible adsorption of the resulting inactive protein, and they are easily fouled by contaminants, i.e., the water molecules that are normally bound at the electrode/electrolyte interface are easily displaced (*cf.* **Figure 2**).

Since the pioneering studies of Rusling and co-workers [4, 5] in the 1990s, the most successful electrode materials for *haem* proteins have been carbon or metal oxides, which bear well-defined natural surface functionalities. Semiconducting metal oxides are often optically

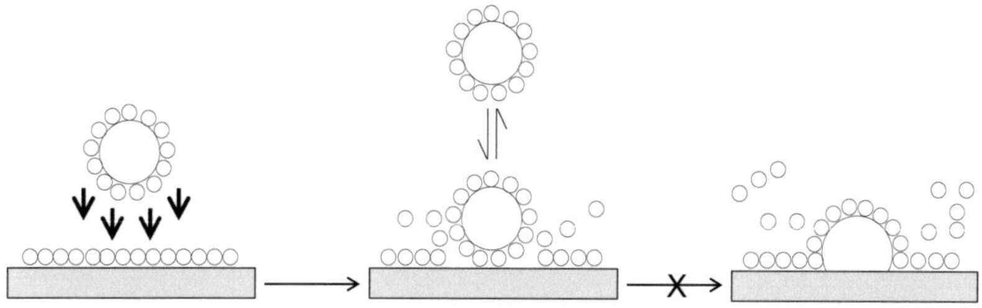

Figure 2. Cartoon illustrating the importance of the protein adsorption process whose conformation may become distorted on interaction with a metallic surface leading to denaturation.

transparent across the visible spectrum and thus provide additional possibilities for spectral studies, e.g., fluorescence and Raman spectroscopies. In the last decade, a few studies have been conducted on mammalian's Hb, for example, Topoglidis et al. [6] reported that titanium oxide and tin oxide allow the reduction of bovine metHb without the addition of any promoters and mediators. Later, Ayato et al. [7, 8] reported that tin-doped indium oxide can induce the electron transfer of the *haem* (Fe^{III}/Fe^{II}) redox center in bovine Hb molecules; they also found that the protein directly adsorbed on the electrode surface was not significantly denatured. More recently, Martinez-Mancera and Hernandez-Lopez [9] reported that thin films of solid solutions like $In_{2-x}Sn_xO_3$ on flat glass substrates can act as both electron acceptors and electron donors, and can be considered a simple model system for mimicking a charge interface of the physiological-binding domain. Herein, the electron transfer properties of the *haem* (Fe^{III}/Fe^{II}) redox center in human Hb molecules were investigated, *in vitro*, on commercial glass/tin-doped indium oxide (ITO) electrodes. Special emphasis is put in theory of cyclic voltammetry and in the Butler-Volmer model, developed by Laviron, for studying the electron transfer between electrode and protein film, the morphological, structural, and surface properties of the electrode, as well as the influence of the physiological milieu that was conditioned into the three-electrode cell system by means of a phosphate-buffered saline (PBS) solution (0.01 mol L^{-1} Na_3PO_4, 0.015 mol L^{-1} NaCl, pH 7.2) and $T = 25°C$. To this chapter, we have added supplementary information. Subsection 2.2.1. A procedure of chromatography in-column, which underlines the importance of preparing and purifying the protein solution. Subsection 2.6. A model of theoretical prediction for determining the point of zero charge of the working electrode.

2. Experimental

2.1. Chemicals

Human hemoglobin (Product No. H7379, $pH_{iep} = 6.87$, MW = 64.5 kDa) and phosphate-buffered saline (PBS) packs (0.01 mol L^{-1} Na_3PO_4, 0.015 mol L^{-1} NaCl, pH 7.2), BupH™ were purchased from Sigma-Aldrich® and Thermo Scientific®, respectively, and used without further purification. Sodium dithionite ($Na_2S_2O_4$), FW = 174.110 g mol^{-1} was purchased from J.T. Baker and used without further purification. BACKBOND spe™, Sephadex® G-25, disposable extraction columns were purchased from J.T. Baker. The concentration of Hb was adjusted to 1×10^{-4} mol Hb L^{-1} using the PBS solution.

2.2. Characterization of the protein by UV-visible spectroscopy

Absorption spectra of human Hb were measured at λ = 200–1000 nm with an UV-Visible spectrophotometer 101 GBS (*Cintra*), using the following parameters: step size = 0.16 nm, scan speed = 400 nm/min, slit width (SW) = 2 nm. The concentration of Hb referred above was estimated by this technique using the following absorptivity value: A_{540} (1%) = 5.97 cm^{-1} [10] (*cf.* **Figure 3**).

2.2.1. Preparation of reduced hemoglobin from oxidized hemoglobin

Reduced hemoglobin can be prepared from oxidized hemoglobin in accordance to the work reported by Dixon and McIntosh [11] with modifications. Briefly, the procedure is as follows: (a) equilibrate a column of Sephadex G-25 (25 × 2.5 cm) with a 20×10^{-3} mol L^{-1} PBS solution, pH 7.0, containing 1×10^{-3} mol L^{-1} EDTA; (b) apply to the column 2 mL of the same buffer to which 1×10^{-3} mol of $Na_2S_2O_4$ have been added, and help it drain into the gel by adding 1 mL of the PBS solution; (c) apply to the column about 10 mL of sample containing oxidized hemoglobin and elute with the PBS solution; (d) saturate the reduced hemoglobin eluent with oxygen gas; and (e) dialyze the oxygenated eluent against an oxygen-saturated PBS solution in order to eliminate any excess of $S_2O_4^{2-}$ and achieve complete conversion to oxyhemoglobin.

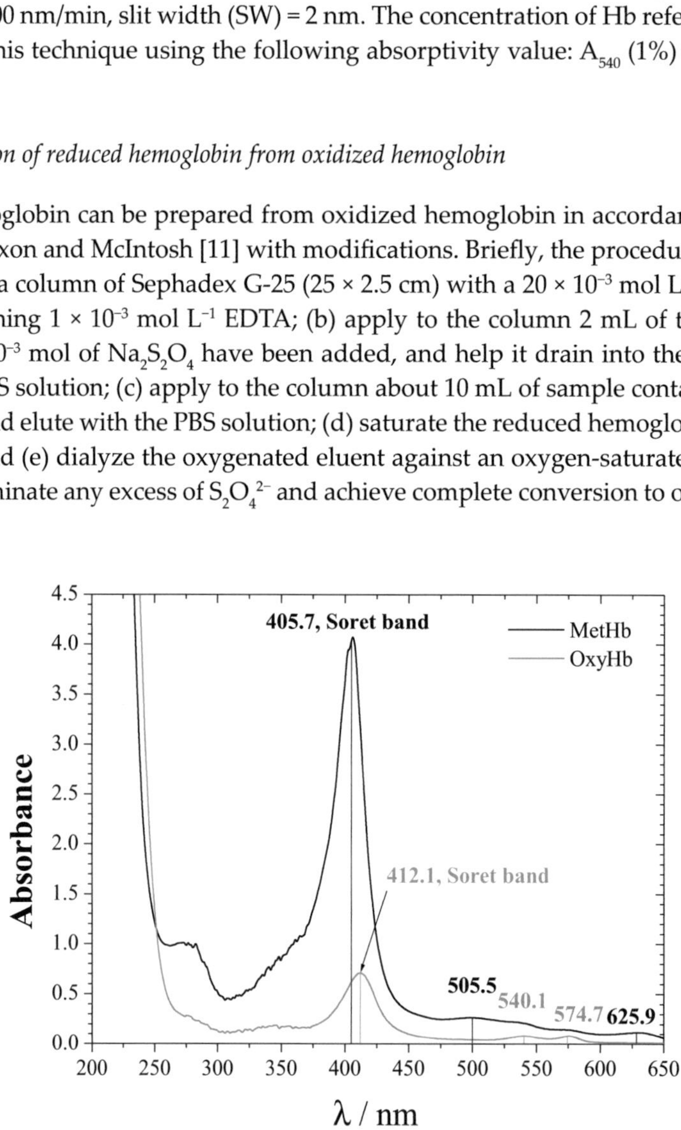

Figure 3. The UV-visible absorption spectra for metHb and oxyHb exhibiting the highly conjugated porphyrin macrocycle with intense features at 405.7 and 412.1 nm (the "Soret" bands), respectively, followed by several weaker absorptions (Q bands) at higher wavelengths (from 450 to 650 nm) [12–14].

http://dx.doi.org/10.5772/67806

2.3. Electrochemical measurement system

Glass/tin-doped indium oxide (ITO) substrates were purchased from *TIRF Technologies, Inc.* The ITO film surface was cleaned according to the following standard procedure [15]: immersion for 15 min each in a series of ultrasonically agitated solvents (acetone, ethanol, water) then for 15 min each in ultrasonically agitated: (a) 2.0% (v/v) phosphate-free detergent solution Hellmanex (*Hellma*™; sonification apparatus Super RK510, *Sonorex*), (b) deionized water type I and (c) ethanol at room temperature. In between the sonification steps, the samples were rinsed in deionized water type I. Finally, the substrates were dried in a stream of nitrogen gas (*Praxair*, 99.999%) until further processing. The electrode potential was controlled with a potentiostat-galvanostat EW-4960 (*Epsilon*™ *BASi*) using a conventional three-electrode cell system supported onto a module C3 (*BASi*). The latter is coupled to a PC/Processor Intel® Celeron, 3.06 GHz. A glass/ITO substrate ($A_g = 1.15$ cm^2) was used as working electrode. A straight platinum wire ($A_g = 0.79$ mm^2) and an electrode of Ag|AgCl in 3 M NaCl solution ($E^{0'} = 0.209$ V vs. SHE at 25°C) were used as counter and reference electrodes, respectively. The cell system was thermostated at 25 ± 0.1°C. Prior to voltammetry, the Hb solution was purged with nitrogen gas (*Praxair*, 99.999%) for at least 30 minutes; then, a nitrogen atmosphere was maintained over the solution during experiments.

2.4. SEM and surface roughness analysis

Scanning electron microscopy (SEM) micrographs were taken with a scanning electron microscope JSM-6510LV (*JEOL*) operated at an accelerating voltage of 15 KV. The superficial characterization of the electrode's roughness was carried out by means of a surface roughness tester HANDYSURF E-35A (*TSK/Carl Zeiss*®) and performing the norm ASME B46.1-2009 Standard.

2.5. XRD analysis

The structural characterization was determined by X-ray powder diffraction (XRD) using a diffractometer D8 Advanced (*Bruker AXS*), using the following parameters: $U = 40$ kV, $I = 35$ mA, Ni-filter, and Cu-Kα radiation, $\lambda = 1.54$ Å. A background diffractogram was subtracted using a glass/tin-doped indium oxide slide as blank. For qualitative analysis, XRD diffractograms were recorded in the interval $10° \leq 2\theta \geq 80°$ at a scan speed of 2°/min.

2.6. Theoretical prediction of the point of zero charge of a glass/ITO electrode

The point of zero charge (PZC) of simple metal oxides can be predicted using an electrostatic model, which takes into account the surface charges originating from the dissociation of amphoteric surface M–OH groups and adsorption of the hydrolysis products of $M^{z+}(OH)^{z-}$ [16]. In this model, a theoretical value of the PZC can be obtained for a given metal oxide by the following equation

$$pH_{pzc} = A - 11.5\left[\frac{z}{R} + 0.0029(\text{CFSE}) + B\right] \quad (1)$$

with

$$R = 2r_O + r_M \quad (2)$$

where z is the ionic charge of species indicated by the subscript, i.e., O = O_2^- and M = cation, r is the ionic radius (r_O = 0.141 nm, r_M = 0.071 nm and 0.081 nm for Sn^{4+} and In^{3+}, respectively), CFSE is a correction factor called crystal field stabilization energy and the constants A and B are parameters that depend on the coordination number of the cation. By virtue of that Sn^{4+} and In^{3+} occupy octahedral interstitial sites in SnO_2 and In_2O_3, the coordination number for these ionic species is 6. Assuming that CFSE is zero in these calculations, A = 18.6 and B = 0 [16]. The predicted PZC values obtained for SnO_2 and In_2O_3 are then $pH_{pzc}(SnO_2) = 5.93$ and $pH_{pzc}(In_2O_3) = 9.37$ [16].

For solid solutions such as ITO, the pH_{pzc} can be considered as a mixture of the constitutive simple oxides and can be calculated by the following equation [17]

$$pH_{pzc} = \Sigma_k s_k pH_{pzc,k} \tag{3}$$

where s_k represents the molar fraction of each constituting oxide at the surface. It can be defined by

$$s_k = \frac{x_k^{2/3}}{\Sigma_k x_k^{2/3}} \tag{4}$$

where x_k is the usual volumetric molar fraction of each constituting oxide.

3. Results and discussion

3.1. Morphological, structural, and electrochemical characterization of a glass/ITO electrode

3.1.1. Morphological and surface roughness analysis

The surface morphology of a pretreated glass/ITO electrode was investigated using SEM (*cf.* **Figure 4**). The micrograph, taken at a 500 μm scale, shows a layer of ITO well defined whose thickness was *ca.* 90 μm. Besides, it is possible to observe a regular, uniform, and flat electrodic surface.

Additionally, the average roughness value obtained in this case was 0.017 μm, which is comparatively lower than the average roughness value obtained for a common glass slide (0.024 μm).

3.1.2. Structural analysis

The structural characterization of a pretreated glass/ITO electrode was investigated using XRD. **Figure 5** shows the XRD pattern of a glass/ITO substrate used like a working electrode. All of the distinct diffraction peaks corresponded to the (211), (222), (400), (440), and (622) reflections of the BCC structure of ITO ($In_{1.94}Sn_{0.06}O_3$) (JCPDS Card File No. 89-4596). Almost all the peaks were very prominent and referred to the cubic rock salt structure of a very crystalline material. Moreover, strong (222) and (400) diffraction peaks are indicative of preferred orientations along the ⟨111⟩ and ⟨100⟩ directions, respectively [18].

http://dx.doi.org/10.5772/67806

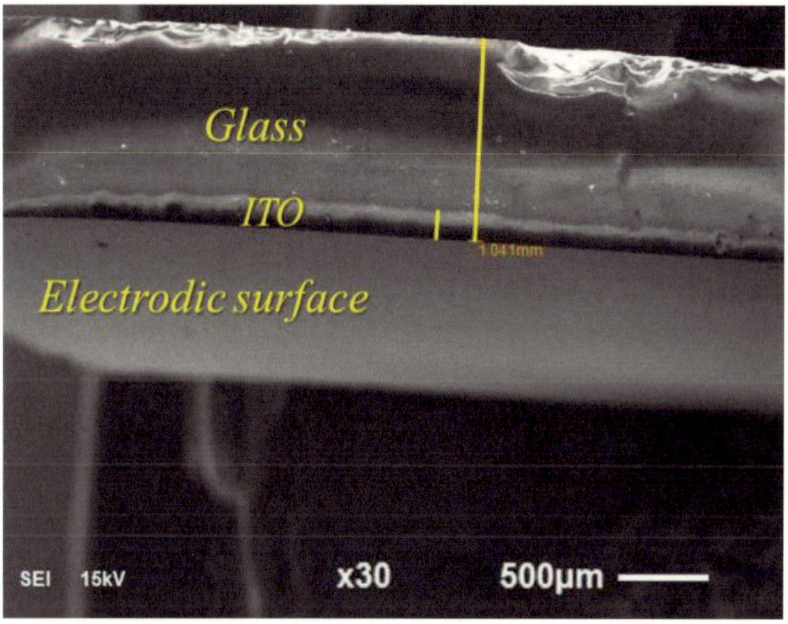

Figure 4. Cross-sectional SEM micrograph of a glass/ITO electrode.

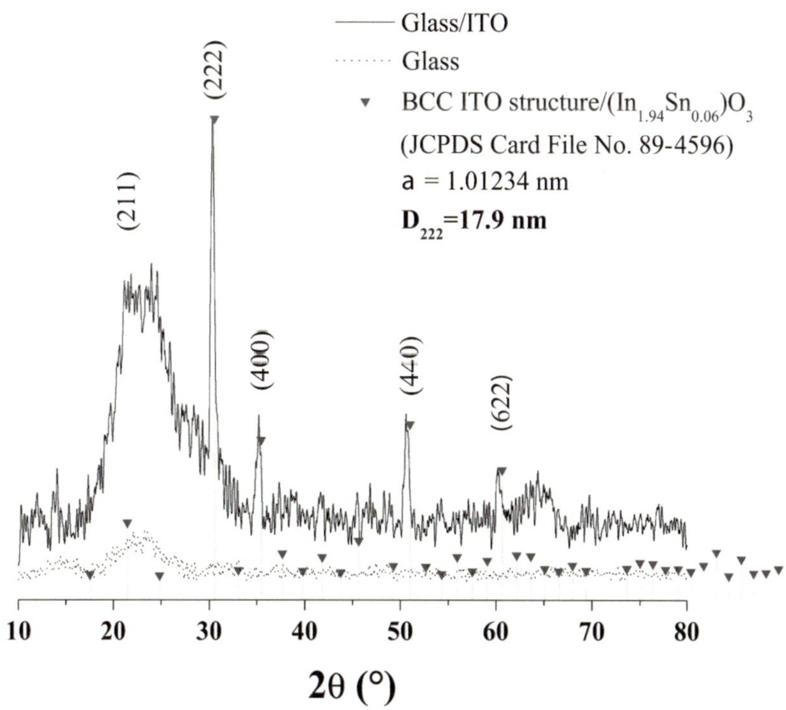

Figure 5. X-ray diffraction pattern of a glass/ITO electrode.

An estimate of the mean crystallite or grain size for a given orientation was determined by using Scherrer's formula [19]:

$$D_{hkl} = \frac{K\lambda}{\beta_{hkl}\cos\theta} \tag{5}$$

where D_{hkl} is the crystallite size (nm), K is a constant (shape factor, about 0.90), λ is the X-ray wavelength (1.54 Å as mentioned before), $\beta_{hkl} = \Delta(2\theta)$ denotes the full width at half maximum (FWHM) or broadening of the diffraction peak (degree), and θ is the diffraction angle (degree). The average D_{hkl} was estimated to be approximately $D_{222} = 17.9$ nm for $2\theta = 30.566°$. It is worth mentioning that the calculated lattice constant a for the glass/ITO substrate using Bragg's equation was $a = 1.01234$ nm, which coincides with the reported value in the standard card.

3.1.3. Electroactive surface area determination

By measuring the peak current in cyclic voltammograms (CVs), the electroactive surface area of a pretreated glass/ITO electrode was determined according to the Randles-Ševčik equation for a reversible electrochemical process under diffusive control:

$$I_{pa} = 0.4463nF\,A_eC_{OX}\sqrt{\frac{nF\nu D}{RT}} \tag{6}$$

where I_{pa} is the anodic peak current (A), n is the number of electrons transferred in the redox reaction, F is Faraday's constant (96,485 C mol^{-1} of electrons), A_e is the electroactive surface area of the electrode (cm^2), C_{OX} is the bulk concentration of an oxidant molecule in the solution (mol cm^{-3}), ν is the scan rate (V s^{-1}), D is the diffusion coefficient of the oxidant molecule in solution, $(6.50 \pm 0.02) \times 10^{-6}$ cm^2 s^{-1}, for hexacyanoferrate (II) in 0.1 mol L^{-1} KCl as supporting electrolyte at 25°C [20], R is the gas universal constant (8.314 J K^{-1} mol^{-1}), and T is the absolute temperature (K).

CVs for 4.0×10^{-3} mol L^{-1} hexacyanoferrate (II) in 0.1 mol L^{-1} KCl were registered to different scan rates (ν = 10, 20, 30, 40, 50, 60, 70, 80, 90, and 100 mV s^{-1}) with the glass/ITO electrode. The peak-to-peak potential separation was constant and linear relationships between the anodic and cathodic peak currents and the square root of the scan rate: $I_{pa} = 0.00101\nu^{1/2} - 5.8550 \times 10^{-7}$, $R^2 = 0.9999$; $-I_{pc} = 0.00101\nu^{1/2} + 3.2078 \times 10^{-7}$, $R^2 = 0.9999$, were achieved. From the slope of these equations, A_e was calculated to be 1.36 cm^2. The roughness factor (ρ) of the GME, which is defined as the ratio (A_e/A_g) [21], was estimated to be 1.18.

3.2. Electrochemical behavior of human hemoglobin molecules

The most popular methods for studying redox enzyme or protein electrochemistry are those based on controlled potential techniques: linear sweep voltammetry (LSV), square wave voltammetry (SWV), and cyclic voltammetry (CV). In the latter, the scan rate, defined as $\nu = \Delta E/\Delta t$, can be varied from less than 1×10^{-3} V s^{-1} to 1×10^{6} V s^{-1} or more, offering a practical timescale window from minutes to microseconds, which makes to this technique very suitable to study interfacial electron transfer kinetics.

Consider the following hypothetical reversible electrochemical reaction: $Ox + ne^- \rightleftharpoons Red$, the interconversion given between oxidized (Ox) and reduced (Red) forms of the protein are fast on the timescale of the voltammogram, as controlled by scan rate. Ideal, reversible voltammograms from a monolayer of electroactive protein on an electrode for a simple electron transfer reaction as it was written above are similar to those of any ultrathin electroactive film. A protein-film voltammetry approach is described in more detail in the following section of this chapter.

3.2.1. *Cyclic voltammetry of thin protein films*

Figure 6 shows the CVs of glass/ITO electrodes in absence and presence of Hb molecules. In **Figure 6a**, a CV recorded in PBS solution alone shows a non-Faradaic current behavior. The electrode had the largest background current in the nonelectrolyte solution which reflected the properties of the electric double layer. The double layer capacitance (C_{dl}) can be estimated by dividing the sum of the anodic and cathodic current with twice the scan rate, i.e., $C_{dl} = (I_{pa} + I_{pc})/2\nu$ [21]. So, the capacitances of the glass/ITO and glass/ITO/Hb electrodes were calculated from **Figure 6a** and **b** as 4.8 and 0.4 μF cm^{-2}, respectively. In **Figure 6b**, a pair of redox peaks, at around −0.117 V in the cathodic scan and at around −0.097 V in the anodic scan, were found in that Hb-containing solution. This fact indicated that a Faradic current was generated over the glass/ITO electrode, which can be adscribed to the *haem* (Fe^{III}/Fe^{II}) redox center in Hb molecules.

Once capacitive effects are counted out, the amount of electrochemically active Hb molecules could be estimated from integration of the charge Q (in C) under each peak in those CVs acquired at slow scan rates (i.e., 0.1, 0.3, or 0.5 V s^{-1}), given by Faraday's law:

$$Q = nFA_e\Gamma_T \tag{7}$$

where Γ_T is total surface concentration of the protein molecule (mol cm^{-2}), A_e is the electroactive surface area of electrode (cm^2), F is Faraday's constant (96,485 C mol^{-1} of electrons) and n is the number of electrons transferred in the redox reaction:

$$\text{Hb: } haem(Fe^{III}) + H_3O^+ + e^- \rightarrow haem(Fe^{III}) \tag{8}$$

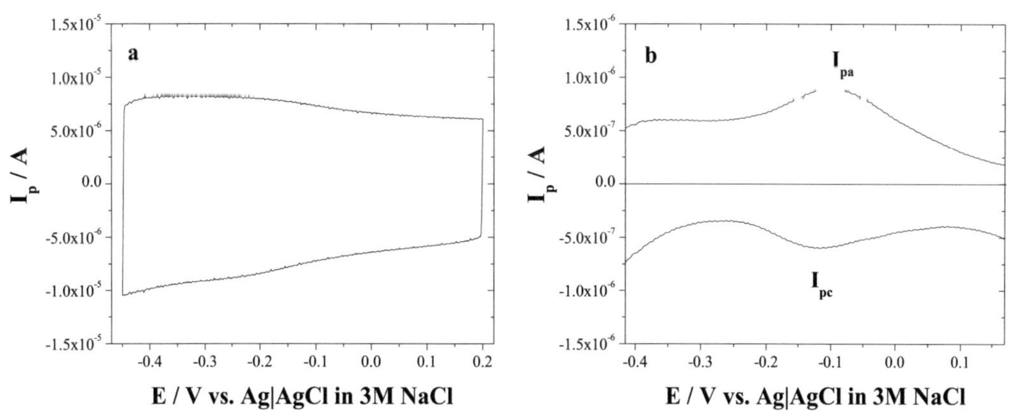

Figure 6. (a) Cyclic voltammograms of: (a) a PBS solution (0.01 mol L^{-1} Na_3PO_4, 0.015 mol L^{-1} NaCl, pH 7.2) and (b) a PBS solution (0.01 mol L^{-1} Na_3PO_4, 0.015 mol L^{-1} NaCl, pH 7.2) containing 1.0 × 10^{-4} mol human Hb L^{-1} after subtraction of (a), as a blank. The experiment was carried out at T = 25°C. Scan rate: 0.5 V s^{-1}.

The total surface concentration of electroactive Hb molecules was estimated to be Γ_T = (4.69 ± 0.52) × 10^{-12} mol cm^{-2}. On the other hand, the theoretical maximum coverage of a protein monolayer on the electrode surface was estimated as 4.89 × 10^{-12} mol cm^{-2}, considering that one human Hb molecule in PBS solution has a Stokes radius of 31.3 Å [22]. These data indicate that $\theta = (\Gamma_T/\Gamma_{T,theo})$ = 0.96 of a protein monolayer was achieved.

Figure 7a shows CVs recorded with different scan rates, from 0.1 to 3.5 V s^{-1}. Nearly symmetric anodic and cathodic peaks were observed; in addition, they have roughly equal heights. The anodic to the cathodic peak potential difference (ΔE_p) was much greater than the ideal value of zero. At the lower scan rates, i.e., 0.1, 0.3, or 0.5 V s^{-1}, the smaller redox peak currents were observed, while CVs with the largest redox peak currents corresponded to those acquired at the fastest scan rate, i.e., 3.5 V s^{-1}.

In such cases, the formal potential $E^{0\prime}$ was taken as the midpoint potential between the oxidation and reduction peaks if there is a small separation between them. Considering this criterium, an $E^{0\prime}$ = −0.107 V (vs. Ag|AgCl in 3 M NaCl solution) was determined at 0.5 V s^{-1}. On the other hand, the anodic and cathodic peak currents, I_{pa} and I_{pc}, increased with increasing scan rates as observed in **Figure 7b**.

These results are characteristic of quasireversible, surface confined electrochemical behavior, in which all electroactive proteins in their *haem* (Fe^{III}) forms are reduced on the forward cathodic scan, and the reduced proteins in their *haem* (Fe^{II}) forms are then fully oxidized to the *haem* (Fe^{III}) forms on the reversed anodic scan.

When the peak currents were plotted against the scan rate, direct linear relationships were obtained, indicating a surface-controlled electrode process. The origin of this process is indicative that the diffusion of H_3O^+ ions toward the electrode surface was very fast. Therefore, the electron process can be expressed as proposed in the redox reaction before [23].

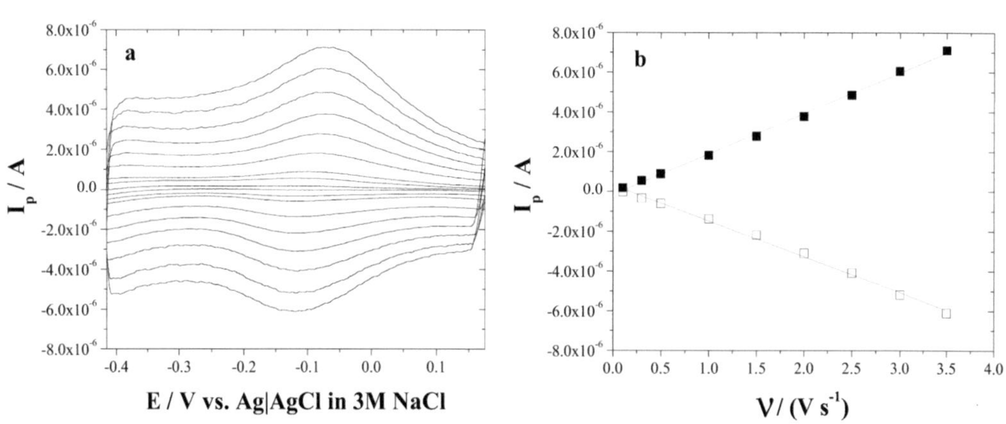

Figure 7. (a) Corrected cyclic voltammograms of a PBS solution (0.01 mol L^{-1} Na_3PO_4, 0.015 mol L^{-1} NaCl, pH 7.2) containing 1.0 × 10^{-4} mol human Hb L^{-1} (T = 25°C) as function of scan rate. Scan rates: 0.1, 0.3, 0.5, 1.0, 1.5, 2.0, 2.5, 3.0, and 3.5 V s^{-1}. (b) Dependence of anodic (■) and cathodic (□) peak currents with scan rate: $I_{pa} = 2.0305 \times 10^{-6}\nu - 1.3809 \times 10^{-7}$, $R^2 = 0.9987$; $-I_{pc} = 1.7896\nu - 3.0310 \times 10^{-7}$, $R^2 = 0.9985$.

The linear regression equations for anodic and cathodic peak currents are as follows: $I_{pa} = 2.0305 \times 10^{-6}\nu - 1.3809 \times 10^{-7}$, $R^2 = 0.9987$; $-I_{pc} = 1.7896\nu - 3.0310 \times 10^{-7}$, $R^2 = 0.9985$.

Linear plots of I_p vs. ν were in good agreement with the following equation [24]:

$$I_p = \frac{n^2F^2A_e\Gamma_T\nu}{4RT} \tag{9}$$

However, their width at half height is nearly 200 mV, much larger than the ideal 90.6/*n* mV at 25°C.

Broadening or narrowing of CV peaks compared to the ideal 90.6/*n* mV at 25°C suggests a breakdown of the ideal model assumptions of no interactions between redox sites that all have the same $E^{0\prime}$. Representative examples arose from studying cytochrome c and myoglobin on Au/alkanethiolate SAMs and OPG/LC surfactants, respectively. Some authors have modeled protein films by LSV, SWV, and CV techniques considering the concepts of spatial distribution of the redox centers, dispersion models of formal potentials ($E^{0\prime}$), and electron transfer rate constants to account for the peak broadening [25–27]. Other factors, e.g., lack of refinement of mathematical algorithms for the extraction of rate constants or improvements to the goodness of fit of SWV data to pulse heights >50 mV, including counterion transport efficiency, could also influence peak widths, but have not been investigated in detail for protein films.

At scan rates <0.5 V s^{-1}, ΔE_p was nearly constant in the films. As the scan rate increased, the peak potentials shifted negatively (*cf.* **Figure 8**). This is consistent with the onset of limiting kinetic effects as scan rates increase.

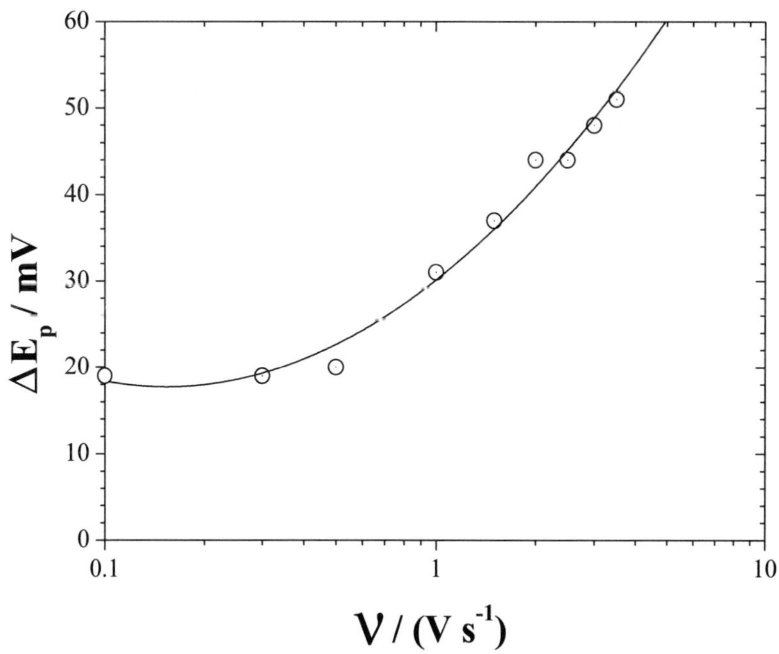

Figure 8. Influence of scan rate on the anodic to the cathodic peak potential difference (ΔE_p) for a PBS solution (0.01 mol L^{-1} Na_3PO_4, 0.015 mol L^{-1} NaCl, pH 7.2) containing 1.0×10^{-4} mol human Hb L^{-1}. The experiment was carried out at T = 25°C.

An increasing ΔE_p as the scan rate is increased for an electroactive thin film suggests kinetic limitations of the electrochemistry [28] and is consistent with predictions of the Butler-Volmer model for electron transfer between an electrode and redox sites in a thin film on an electrode. Possible causes could be attributed to: (a) slow electron transfer between electrode and redox centers, (b) slow transport of charge within the film limited by electron or counterion transport, (c) uncompensated voltage drop within the film, and (d) structural reorganization of the protein accompanying the redox reactions.

When $n\Delta E_p < 200$ mV, the surface electron transfer rate constant (k_s) of the adsorbed Hb on the glass/ITO electrode can be estimated according to Laviron's equation for quasi-reversible thin-layer electrochemistry [29]:

$$\text{Log}\, k_s = \alpha\, \text{Log}\,(1-\alpha) + (1-\alpha)\,\text{Log}\,\alpha - \text{Log}\frac{RT}{nF\nu} - \alpha(1-\alpha)\frac{nF\Delta E_p}{2.3RT} \qquad (10)$$

Our experimental results showed that the scan rate in the range 0.1–3.5 V s^{-1} did not affect the k_s value, because $n\Delta E_p < 200$ mV. Assuming a charge-transfer coefficient α of 0.5, the k_s of the adsorbed Hb thin film on the glass/ITO electrode was 8.01 s^{-1} at the onset of limiting kinetic effects (500 mV s^{-1}). This value is significantly higher than other previously reported values in the literature for different hemoglobin species and electrode materials (*cf.* **Table 1**). For comparison with data on bare or mediator-coated electrodes, k_s was converted to the standard heterogeneous rate constant ($k^{0\prime}$) by using $k^{0\prime} = k_s\ d$, where d is the film thickness [29]. This fact could be attributed to the morphological and structural properties shown by the electrode, i.e., surface roughness, crystallinity, and hydrophilicity [30], as well as the influence of the physiological milieu that was conditioned into the three-electrode cell system, charging positively/negatively to the working electrode (*cf.* Section 2.6 and Ref. [31]) and negatively to the protein [32]. All these factors played an important role in providing a more favorable micro-environment for the protein.

pH	Sample | electrode	E$^{0\prime}$/mV (SHE)	$k^{0\prime}$/cm s^{-1}	References
5.5	^{b}Hb-DDAB-Nafion | edge-plane PG	80	5.7 d^*	[4]
5.5	^{h}Hb-DDAB | edge-plane PG	84	2.7 d^*	[5]
5.5	^{h}Hb in solution | edge-plane PG	Not detected	Not detected	op. cit. [5]
7.0	^{b}Hb in solution | Pt + MB	145	2.0×10^{-4}	op. cit. [5]
7.0	^{b}Hb in solution | Pt + Azure A	180	3.5×10^{-6}	op. cit. [5]
7.0	^{b}Hb in solution | Pt + BCG	184	2.0×10^{-7}	op. cit. [5]
7.0	^{b}Hb in solution | SnO2	−215	0.53 d^*	[6]
7.4	^{b}Hb in solution | $In_{2-x}Sn_xO_3$	−112	Not determined	[7, 8]
7.2	^{h}Hb in solution | $In_{1.94}Sn_{0.06}O_3$	102	8.01 d^*	[9]

DDAB = Didodecyldimethylammonium bromide, PG = pyrolytic graphite, MB = methylene blue, BCG = brilliant cresyl green.

*For comparison with CV data where diffusion control pertains, k_s was estimated from the standard heterogeneous rate constant ($k^{0\prime}$) by using $k^{0\prime} = k_s \cdot d$, where d is the film thickness [29].

Table 1. Electrochemical parameters for different mammalian (superscripts: b-bovine, h-human) hemoglobin species at 25°C.

http://dx.doi.org/10.5772/67806

As indicated in Section 1, Introduction, to facilitate the electron communication between the prosthetic group of *haem* proteins and an electrode due to misalignment of the *haem* (Fe^{III}/Fe^{II}) redox center is a difficult task but could help to advance understanding on biological electron transfer. Techniques such as FT-IR spectroscopy, NMR, ESR anisotropy, polarized reflectance FT-IR, circular dichroism, and calorimetry could yield detailed information on the secondary structure of the protein in regards to its redox state as well as to give some notion of order and specific orientation.

4. Conclusions

In this study, we clearly demonstrated that human Hb molecules directly physisorbed on glass/tin-doped indium oxide substrates exhibited direct electron transfer (DET) in PBS (0.01 mol L^{-1} Na_3PO_4, 0.015 mol L^{-1} NaCl, $pH = 7.2$) solution and T = 25°C.

The experimental results suggest that acid-base equilibria and the water molecule coordinated to the *haem* group as the sixth ligand might play an important role in the electron transfer process between human hemoglobin and very crystalline and hydrophilic tin-doped indium oxide electrodes.

Acknowledgments

The authors would like to acknowledge to José Germán Flores-López from Departamento de Servicios Tecnológicos (CIDETEQ, S.C.) for his technical assistance with the SEM microscope and surface roughness tester and to the National Council for Science and Technology (CONACyT) for its financial support on this research project (FOMIX-QRO-2007-C01, Project Nr. 78809; CB-2008-C01, Project Nr. 101701; Salud-2009-01, Project Nr. 114166).

Author details

Flavio Dolores Martínez-Mancera and José Luis Hernández-López*

*Address all correspondence to: jhernandez@cideteq.mx

Center of Research and Technological Development in Electrochemistry, Parque Tecnológico Querétaro S/N, Pedro Escobedo, Mexico

References

[1] Perutz MF, Wilkinson AJ, Paoli M, Dodson GG. The stereochemical mechanism of the cooperative effects in hemoglobin revisited. Annual Review of Biophysics and Biomolecular Structure. 1998;**27**:1–34. DOI: 10.1146/annurev.biophys.27.1.1

[2] Schumacher MA, Dixon MM, Kluger R, Jones RT, Brennan RG. Allosteric transition intermediates modelled by crosslinked haemoglobins. Nature. 1995;**375**(6526):84–87. DOI: 10.1038/375084a0

[3] Armstrong FA. Voltammetry of proteins. Wilson, G.S., editors. Bioelectrochemistry. Weinheim, Germany: Wiley-VCH Verlag GmbH; 2002. pp. 11–29

[4] Huang Q, Lu Z, Rusling JF. Composite films of surfactants, nafion, and proteins with electrochemical and enzyme activity. Langmuir. 1996;**12**(22):5472–5480. DOI: 10.1021/la9603784

[5] Lu Z, Huang Q, Rusling JF. Films of hemoglobin and didodecyldimethylammonium bromide with enhanced electron transfer rates. Journal of Electroanalytical Chemistry. 1997;**423**(1–2):59–66. DOI: 10.1016/S0022-0728(96)04843-7

[6] Topoglidis E, Astuti Y, Duriaux F, Grätzel M, Durrant JR. Direct electrochemistry and nitric oxide interaction of heme proteins adsorbed on nanocrystalline tin oxide electrodes. Langmuir. 2003;**19**(17):6894–6900. DOI: 10.1021/la034466h

[7] Ayato Y, Itahashi T, Matsuda N. Direct electron transfer of hemoglobin molecules on bare ITO electrodes. Chemistry Letters. 2007;**36**(3):406–407. DOI: 10.1246/cl.2007.406

[8] Ayato Y, Takaktsu A, Kato K. Matsuda, N. Direct electrochemistry of hemoglobin molecules adsorbed on bare indium tin oxide electrode surfaces. Japanese Journal of Applied Physics. 2008;**47**(2S):1333–1336. DOI: 10.1143/JJAP.47.1333

[9] Martínez-Mancera FD, Hernández-López JL. In vitro observation of direct electron transfer of human haemoglobin molecules on glass/tin-doped indium oxide electrodes. Journal of the Mexican Chemical Society. 2015;**59**(4):302–307

[10] Li T-K, Johnson BP. Optically active heme bands of hemoglobin and methemoglobin derivatives. Correlation with absorption and magnetic properties. Biochemistry. 1969;**8**(9):3638–3643. DOI: 10.1021/bi00837a021

[11] Dixon HBF, McIntosh R. Reduction of methaemoglobin in haemoglobin samples using gel filtration for continuous removal of reaction products. Nature. 1967;**213**(5074):399–400. DOI: 10.1038/213399a0

[12] Smith KM., editors. Porphyrins and Metalloporphyrins. Amsterdam, The Netherlands: Elsevier; 1975. p. 910

[13] Boulton M, Rozanowska M, Rozanowski B. Retinal photodamage. Journal of Photochemistry and Photobiology B: Biology. 2001;**64**(2–3):144–161. DOI: 10.1016/S1011-1344(01)00227-5

[14] Sagun EI, Zenkevich EI, Knyukshto VN, Shulga AM, Starukhin DA, Borczyskowski CV. Interaction of multiporphyrin systems with molecular oxygen in liquid solutions: Extra-ligation and screening effects. Chemical Physics. 2002;**275**(1–3):211–230. DOI: 10.1016/S0301-0104(01)00517-1

[15] Hernandez-Lopez JL. Water-soluble dendrimers: A novel hierarchical concept for the preparation and study of supramolecular interface based films [dissertation]. Mainz,

Germany: Fachbereich Chemie und Pharmazie der Johannes Gutenberg-Universität; 2003. p. 297

[16] Parks GA. The isoelectric points of solid oxides, solid hydroxides, and aqueous hydroxo complex systems. Chemical Reviews. 1965;**65**(2):177–198. DOI: 10.1021/cr60234a002

[17] Carre A, Roger F, Varinot C. Study of acid/base properties of oxide, oxide glass, and glass-ceramic surfaces. Journal of Colloid Interface Science. 1992;**154**(1):174–183. DOI: 10.1016/0021-9797(92)90090-9

[18] Meng L-J, dos Santos MP. Properties of indium tin oxide films prepared by rf reactive magnetron sputtering at different substrate temperature. Thin Solid Films. 1998;**322**(1–2):56–62. DOI: 10.1016/S0040-6090(97)00939-5

[19] Cullity BD. Elements of X-ray Diffraction. 2nd ed. Massachusetts, USA: Addison-Wesley, Publishing Company; 1978. p. 569

[20] Stackelberg MV, Pilgram M, Toome V. Bestimmung von Diffusionskoeffizienten einiger Ionen in wäßriger Lösung in Gegenwart von Fremdelektrolyten. I. Berichte der Bunsengesellschaft für Physikalische Chemie. 1953;**57**(5):342–350. DOI: 10.1002/bbpc.195300073

[21] Bard AJ, Faulkner LR. Electrochemical Methods: Fundamentals and Applications. 2nd ed. New York, USA: John Wiley & Sons, Inc.; 2001. p. 850

[22] Gros G. Concentration dependence of the self-diffusion of human and Lumbricus terrestris hemoglobin. Biophysical Journal. 1978;**22**(3):453–468. DOI: 10.1016/S0006-3495(78)85499-X

[23] Wang C, Yang C, Song Y, Gao W, Xia X. Adsorption and direct electron transfer from hemoglobin into a three-dimensionally ordered macroporous gold film. Advanced Functional Materials. 2005;**15**(8):1267–1275. DOI: 10.1002/adfm.200500048

[24] Rusling JF, Zhang Z. Thin films on electrodes for direct protein electron transfer. In: Nalwa, R.W., editors. Handbook of Surfaces and Interfaces of Materials. San Diego, USA: Academic Press, Inc.; 2001. pp. 33–71. DOI: 10.1016/B978-012513910-6/50059-1

[25] Nahir TM, Clark RA, Bowden EF. Linear sweep voltammetry of irreversible electron transfer in surface-confined species using the Marcus theory. Analytical Chemistry. 1994;**66**(15):2595–2598. DOI: 10.1021/ac00087a027

[26] Zhang Z, Rusling JF. Electron transfer between myoglobin and electrodes in thin films of phosphatidylcholines and dihexadecylphosphate. Biophysical Chemistry. 1997;**63**(2–3): 133–146. DOI: 10.1016/S0301-4622(96)02216-8

[27] Nassar A-EF, Zhang Z, Hu N, Rusling JF, Kumosinski TF. Proton-coupled electron transfer from electrodes to myoglobin in ordered biomembrane-like films. The Journal of Physical Chemistry B. 1997;**101**(12):2224–2231. DOI: 10.1021/jp962896t

[28] Murray RW. Chemically modified electrodes. In: Bard, A.J., editors. Electroanalytical Chemistry. New York, USA: Marcel Dekker, Inc.; 1984. pp. 191–368

[29] Laviron E. General expression of the linear potential sweep voltammogram in the case of diffusionless electrochemical systems. Journal of Electroanalytical Chemistry and Interfacial Electrochemistry. 1979;**101**(1):19–28. DOI: 10.1016/S0022-0728(79)80075-3

[30] Li C-Z, Liu G, Prabhulkar S. Comparison of kinetics of hemoglobin electron transfer in solution and immobilized on electrode surface. American Journal of Biomedical Sciences. 2009;**1**(4):303–311. DOI: 10.5099/aj090400303

[31] Lin X-Y, Farhi E, Arribart H. Determination of the isoelectric point of planar oxide surfaces by a particle adhesion method. The Journal of Adhesion. 1995;**51**(1–4):181–189. DOI: 10.1080/00218469508009997

[32] Conway-Jacobs A, Lewin LM. Isoelectric focusing in acrylamide gels: Use of amphoteric dyes as internal markers for determination of isoelectric points. Analytical Biochemistry. 1971;**43**(2):394–400. DOI: 10.1016/0003-2697(71)90269-7

Chapter 4

Advances and Trends in Voltammetric Analysis of Dyes

Felipe Fantinato Hudari,

Michelle Fernanda Brugnera and

Maria Valnice Boldrin Zanoni

Additional information is available at the end of the chapter

http://dx.doi.org/10.5772/67945

Abstract

Since 1856 when W. H. Perkin synthesized the first synthetic dye (Mauveine), a wide variety of colors and shades are produced and used in several commercial products. The occurrence in water and wastewater has gained controversy regarding their toxicity and mutagenicity and it has been regulation by several regulatory agencies. Thus, analytical methods able to determine these colorings in several matrices with high sensitive and robust enough are relevant. Among several analytical methods, the use of electroanalytical methods, especially the voltammetric techniques, are of great interest due to the high selectivity, sensitivity, use of low quantity of sample, little or without sample treatment, and low waste generation, which contributes to reduced environmental impact. Over the past decades, the technical based on current-potential curves by using of static electrodes have gained considerable progress, as minimizing the effect of capacitive current and the possibility of pre-concentration of the analyte at the electrode surface, which has reflected in lower detection levels. The present work gives an overview about the analytical methods available in literature focusing on electroanalysis of dyes by using voltammetric techniques. The advances of the electroanalytical techniques and the use of different modifiers to increase sensitivity and selectivity are reviewed.

Keywords: voltammetry trends, dyes analysis, electroanalytical methods, dye determination

1. Introduction

Since 1856, when W.H. Perkin synthesized the first synthetic dye (mauveine), a wide variety of colors and shades has been produced and used in several commercial products, mainly in the industries of textiles, cosmetics, food, and others [1–4]. The dyes present complex organic structures with several chromophoric centers based on functional groups, such as azo, anthraquinone, polymethine, nitro, nitroso, aryl methyl, xanthine, coumarin, and others. They also have some particular physicochemical properties that are essential for their attachment in specific types of natural fibers, such as cotton, silk, leather, and hair and synthetic fibers, such as polyamide, polyester, and cellulose acetate. Thus, they can be commercially classified as due chromophoric group or due fixation to the fiber and can also be assigned as reactive, direct, acid, vat, sulfur, dispersed, premetalized, optical brighteners, and so on.

The use and release of dyes in the environment has received great attention since approximately 9% (40,000 tons) of the dyes produced worldwide (450,000 tons) are discharged in to textile wastewater [5, 6]. Their occurrence in water and wastewater has gained controversy regarding their toxicity and mutagenicity, and it has been regulated by several regulatory agencies [7]. Besides that, their addition in food has obeyed rigorous control, and a little amount of dye is allowed for this strict end. Thus, analytical methods able to determine these colorings in surface water, commercial formulation, industrial effluents, and food are of great interest. For monitoring in environmental samples, where the dye is much diluted, is required very sensitive and robust methods as well in food samples that usually are based on complexes matrices. In this context, several analytical methods are described in literature and are compiled here. The most popular method is the UV–visible (UV–vis) spectrophotometry, and it is based on the presence of chromophoric groups responsible for the color of the dye solution [8–10]. However, due to its medium sensitivity coupled with matrix interference and band overlapping in simultaneous measurements, this technique is losing space. The development of methods for the detection of dyes using chromatographic techniques has also been exploited, particularly due to low levels of detection and high types of available detectors [11–14]. Nevertheless, while chromatography-based methods are effective for the detection and quantification of the dyes in the wide range of matrices, such methods require the use of a large amount of organic solvents and a laborious sample preparation. In this context, electrochemical techniques, especially the voltammetric techniques, have been used as alternative methods due to the high selectivity, sensitivity, low cost, use of low-quantity sample, little or no sample treatment, and low waste generation, which contribute to reduced environmental impact.

The voltammetric techniques gained notoriety in the early 1920s, when Jaroslav Heyrovsky developed the polarography based on the use of dropping mercury electrode [15]. Over the past decades, the technical based on current-potential curves (polarography and voltammetry) by using static electrodes have gained considerable progress, as minimizing the effect of capacitive current (techniques of differential pulse voltammetry and square wave voltammetry) and the possibility of preconcentration of the analyte at the electrode surface, which has reflected in lower detection levels. Furthermore, the use of stationary solid electrodes (electrodes as gold, platinum, and carbon) has shown improvements, because there is the possibility of modifying

http://dx.doi.org/10.5772/67945

the surface of these electrodes with high diversities of materials, such as metal nanoparticles, polyaminoacids, carbon nanotubes, graphene, and imprinted molecular polymer, which improved considerably the sensitivity and/or selectivity of the electroanalytical method [16–18].

The present work gives an overview about the analytical methods available in literature, focusing on electroanalysis of dyes by using voltammetric techniques.

2. Electroanalysis of food dyes

Food additives are substances (or mixtures) that are added during the process of food manufacturing, processing, or packaging with the purpose to prevent changes and/or to confer, intensify, and maintain color, aroma, taste, or any other action required to improve the quality or aspect of the food [19–21]. Among the food additives used, we can highlight the color additives, usually added only to make them more attractive and tasty to the consumers [20, 22].

Until the middle of the nineteenth century, all the coloration used in dyes came from extracts of animals or vegetables [20, 23]. But currently, natural dyes have been replaced by synthetic dyes, because they present better stability, uniformity, and tinctorial power [6, 24, 25]. As a reflection of this progress, at the end of the nineteenth century, more than 90 dyes were used by the food industries. In 1906, started in the USA, a great concern and the first legislation was imposed to the control the use of food colorants, in which only seven dyes were authorized [26].

Color additives can be classified in different ways. In Brazil, a simple one is stipulated by the *Agência Nacional da Vigilância sanitária* (ANVISA) in the resolution of the *Nacional de Normas e Padrões para Alimentos* (CNNPA) n° 44, of 1977. These agencies establish that food colorings can be classified as natural organic dyes (derived from vegetables or animals), artificial dyes (synthetic organic dyes), synthetic organic dyes identical to natural dyes (synthetic organic dyes whose chemical structure is derived from natural organic dyes), inorganic dyes (obtained from mineral substances), and caramel dyes (natural dyes obtained by heating sugars) [23, 27]. **Figure 1** exemplifies the distribution of the dyes used in food (food and drink) products in the world, which shows the supreme use of synthetic additives by food industries.

Due to the inefficiency of resolutions, control, and monitoring, many illegal dyes were used in food products, which resulted in cases of allergic reactions and even deaths, as reported in 1860 after two people consumed a dessert that contained copper arsenate [23]. In India, studies indicated that 61.6% of the analyzed foods presented dyes not allowed in the country [28]. This occurs because the synthesis in most cases is complex and the purification requires time and money.

Food legislation is a very important factor in ensuring the quality of food. The first supervisory agency was created in the USA and became known as the 1906 FD&C Act [26]. Nowadays, there are several agencies that dictate and supervise the dye additives allowed, not only to maintain the good quality of food but also to preserve the health of the people, since research indicates that certain substances with dyeing power have great mutagenic potential besides adverse reactions. In the USA, all color additives are regulated by the Federal Food, Drug, and Cosmetic Act (FD&C) [29, 30]. However, the use of additives must be approved by the USA Food and Drug Administration (FDA) by a color petition process and listed in Title 21 of the Code of Federal Regulations (CFR, 2014) [23, 30, 31]. In the European Union (EU), the first legislation was created in 1950 [26]. Currently, food additives are controlled by EU Regulation (EC) No1333/2008 and food additives are divided into 20 groups, according to their functionality [30]. Among them, 43 colorants are permitted as food additives, of which 17 are synthetic and 26 are natural [26].

Table 1 shows some types of synthetic color additives commonly used in food and whether they are recognized or not by the European Union (EU), US Food and Drug Administration (FDA), *Agência Nacional da Vigilância sanitária* (ANVISA), and World Health Organization (WHO).

The search for highly sensitive, efficient, and rapid methods of analysis of these additives has been growing. This is justified by a great demand for coloring additives to be used in food. Therefore, the inspection agencies need to specify the types of substances that can be used and chosen one that is not harmful for the human health. In addition, the presence of food dyes has also been reported in wastewater, and there is a demand for analytical methods applicable in environmental matrices.

The first works for the detection of food dyes using electroanalytical methods are dated to 1979, and they were based on the use of mercury electrodes [33, 34]. Fogg and Yoo [33] have used differential-pulse polarography (DPP) for the determination of mixtures of Tartrazine-Sunset Yellow FCF, Tartrazine-Green S, and Amaranth-Green S in soft drinks. In another work, FOGG et al. [35] used differential-pulse adsorptive stripping voltammograms (DPASV) for the determination of 13 food dyes assigned as Amaranth, Carmoisine, Ponceau 4R, Red 2G, Erythrosine, Sunset Yellow FCF, Tartrazine, Quinoline Yellow, Green S, Indigo carmine, Patent Blue V, Brilliant Blue FCF, and Chocolate Brown HT. For Amaranth, Carmoisine, Ponceau 4R, Red 2G, Sunset Yellow FCF, Tartrazine, and Chocolate Brown HT, the signals were attributed to the reduction of the azo to hydrazo group. After optimization for each analyte individually, the sensitivities for food dyes were in the range of 5.5–38 and 120–1500 mA L mol^{-1} by using dropping mercury electrode (DME) and hanging mercury drop electrode (HMDE), respectively.

Dominguez and collaborators [36] have demonstrated that Sunset Yellow FCF and Tartrazine can be determined in soft drinks sample by DME and DPP techniques. The sensitivities of the

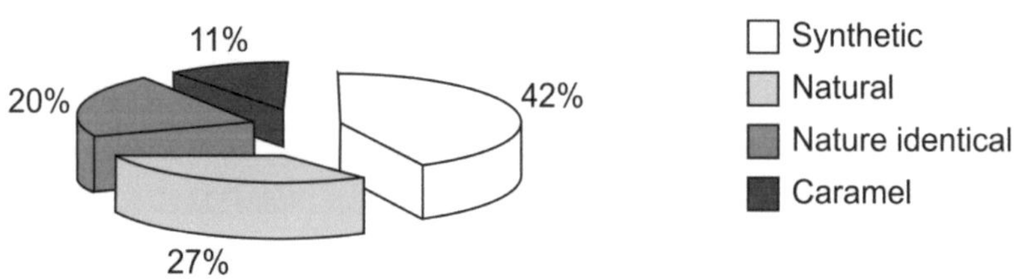

Figure 1. Distribution of the types of coloring additives used in food in the 1992 world market [23].

http://dx.doi.org/10.5772/67945

Color additives	Permission to use in food				
	Synonymous	EU	FDA	ANVISA	WHO
Amaranth 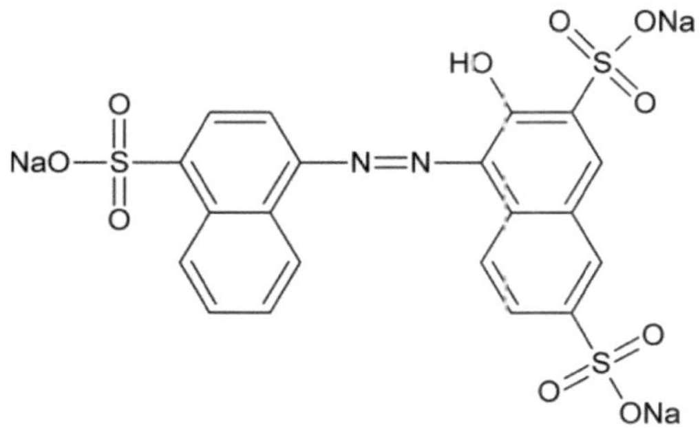	Acid Red 27 Azorubin SFD&C Red Dye No. 2	No	No	Yes	Yes
Allura red 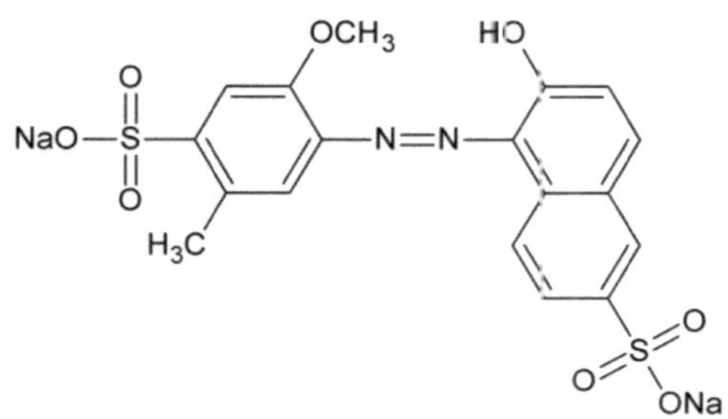	Disodium 6-hydroxy-5-[(2-methoxy-5-methyl-4-sulfophenyl) azo]-2-naphthalenesulfonate FD&C red No. 40	No	Yes	Yes	Yes

Color additives	Permission to use in food				
	Synonymous	EU	FDA	ANVISA	WHO
Brilliant black BN 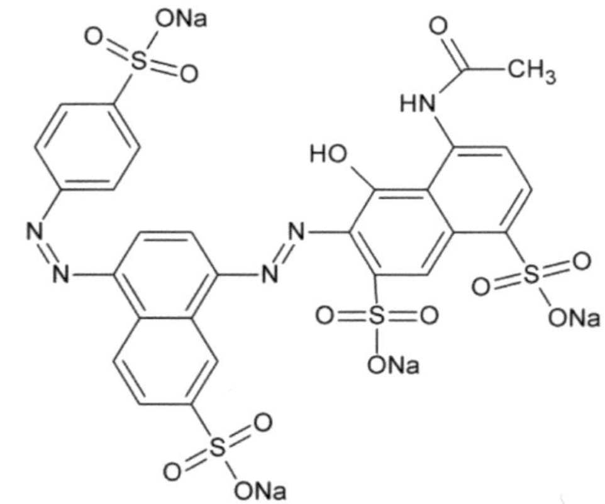	Food Black 1	No	No	No	Yes
Sunset yellow 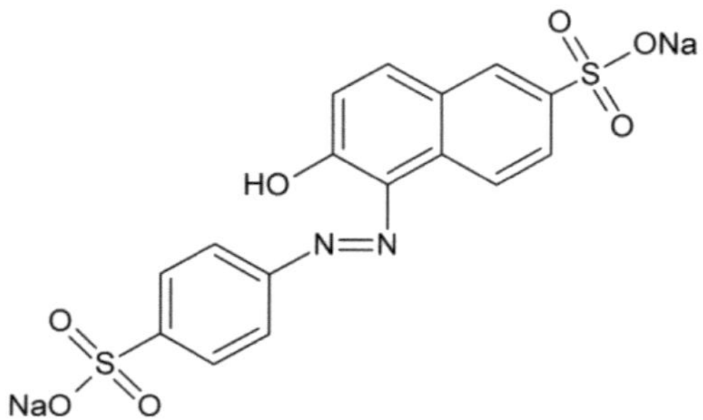	Food Yellow 3 Orange Yellow S FD&C Yellow No. 6	Yes	Yes	Yes	Yes

http://dx.doi.org/10.5772/67945

Color additives	Permission to use in food				
	Synonymous	EU	FDA	ANVISA	WHO
Carmoisine 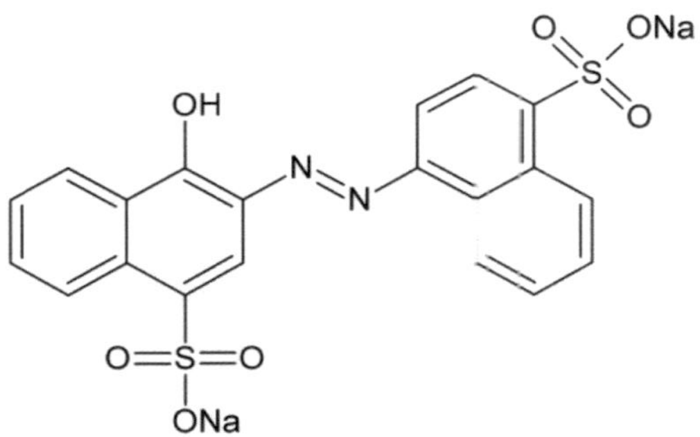	Acid Red 14 Chromotrope FB Disodium 4-hydroxy-3-[(4-sulfo-1-naphthalenyl)azo]-1-naphthalenesulfonate Mordant Blue 79	Yes	No	Yes	Yes
Tartrazine 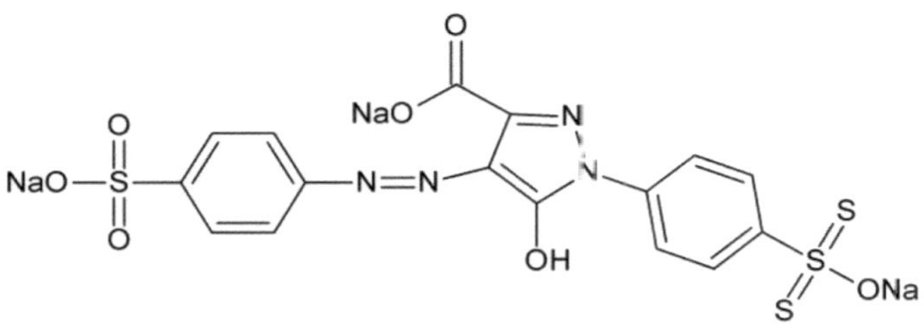	Acid Yellow 23 FD&C Yellow No. 5	Yes	Yes	Yes	Yes

Color additives	Permission to use in food				
	Synonymous	EU	FDA	ANVISA	WHO
Ponceau 4R 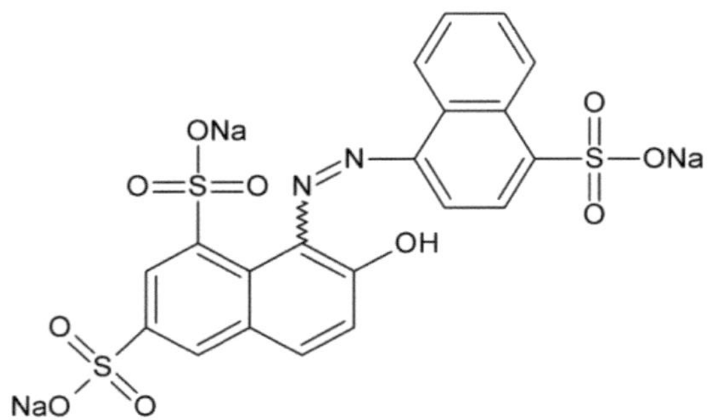	Acid Red 18 New Coccine Ponceau 4 R	Yes	No	Yes	Yes
Brilliant blue FCF 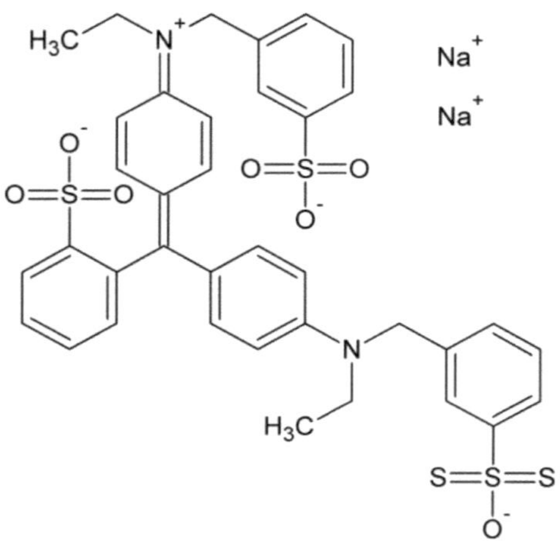	Acid Blue 9 Alphazurine FG E133 Erioglaucine disodium salt FD&C BLUE No. 1	No	Yes	Yes	Yes

http://dx.doi.org/10.5772/67945

Color additives	Permission to use in food				
	Synonymous	EU	FDA	ANVISA	WHO
Brown HT 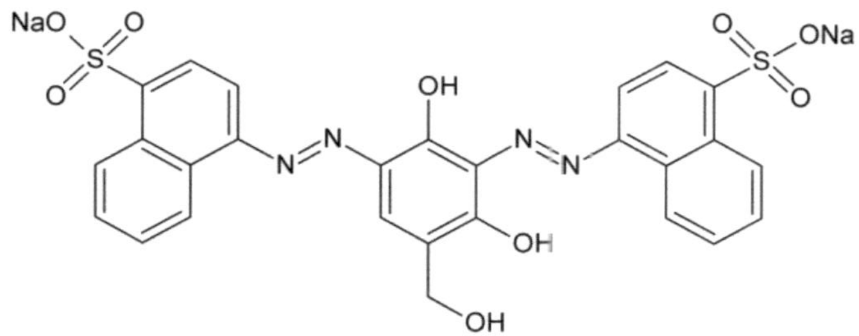	Chocolate Brown HT Food Brown 3 C.I. 20285	No	No	No	Yes
Red 2G 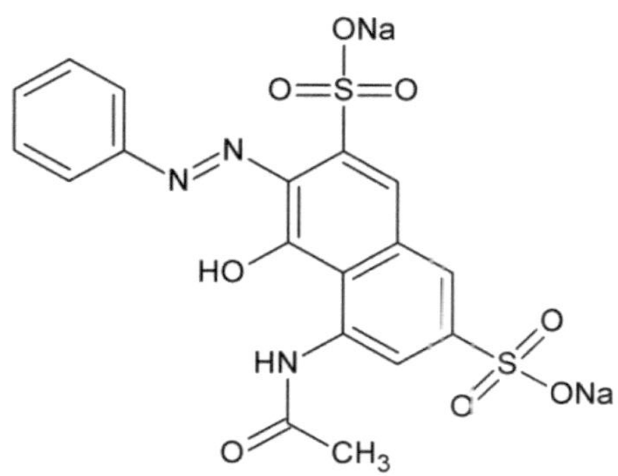	Acid Red 1, Amidonaphthol Red G, Azophloxine	Yes	No	No	

Color additives	Permission to use in food				
	Synonymous	EU	FDA	ANVISA	WHO
Fast green FCF 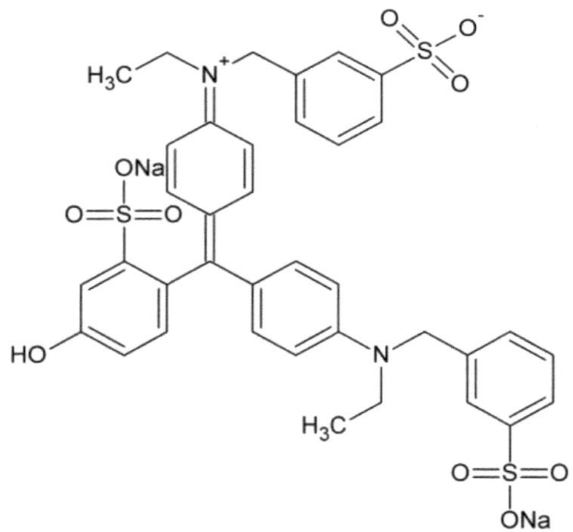	Food green 3 FD&C Green No. 3	No	Yes	Yes	Yes

http://dx.doi.org/10.5772/67945

Color additives	Permission to use in food				
	Synonymous	EU	FDA	ANVISA	WHO
Patent blue V 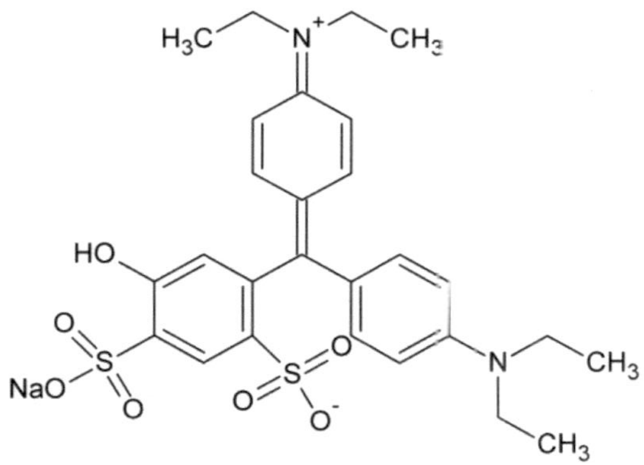	Acid blue 3 sodium salt Food Blue 5 sodium salt	Yes	No	Yes	Yes
Indigotine 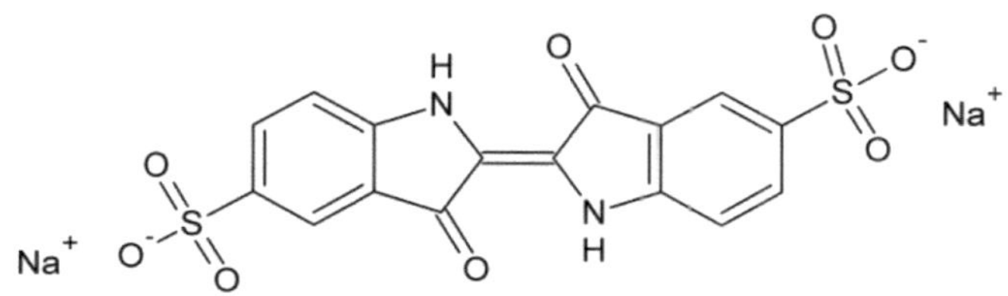	Indigo carmine, Acid Blue 74, Indigo-5,5′-disulfonic acid disodium salt Indigocarmine FD&C Blue No. 2	Yes	Yes	Yes	Yes

Color additives	Permission to use in food				
	Synonymous	**EU**	**FDA**	**ANVISA**	**WHO**
Erythrosine 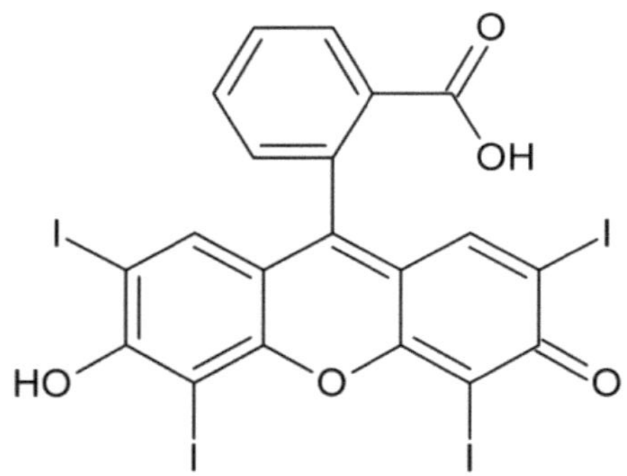	FD&C Red No. 3	Yes	Yes	Yes	Yes
Fast red E 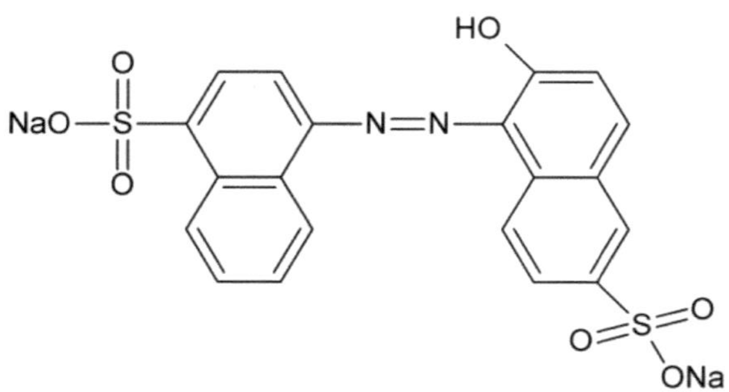	CI Food Red 4 CI (1975) No. 16045	No	No	No	Yes

Table 1. Synthetic color additives used in food in EU, FDA (USA), ANVISA, and WHO (adapted from Refs. [26, 32]).

proposed methods were 4.56×10^4 (±0.03) and 1.98×10^4 (±0.02) μA L mol^{-1} and the detection limits were 1.3×10^{-8} and 3.0×10^{-8} mol L^{-1} for Sunset Yellow FCF and Tartrazine, respectively. In the work of ALBA and co-authors [37], a method was developed for the determination of the synthetic food colorants Tartrazine, Allura Red, and Sunset Yellow by DPP. Using as a response, for the three analytes, the signal intensity related to the reduction of the azo to hydrazo group on the surface of a DME, linear relationships were obtained between 0.050 and 7.5, 0.050 and 7.5, and 0.050 and 10 μmol L^{-1}, and the detection limits 0.013, 0.020, and 0.011 μmol L^{-1} for Allura Red, Tartrazine, and Sunset Yellow, respectively. Combeau et al. [38] proposed a method for the determination of Azorubin, Allura Red, and Ponceau 4R in soft drinks by DPP. Reductions of the azo groups (for all dyes) were promoted under the surface of a DME. Using as electrolyte 0.1 mol L^{-1} KCl, the detection limits were 22, 50 and 44 μg L^{-1} for Azorubin, Allura Red, and Ponceau, respectively. Finally, the application in the food dye sample showed low values of standard deviation in relation to the amount added and found for all analytes.

Although most studies are based on the reduction of food dye molecule as a basis for its monitoring, some chapters describe the use of electrochemical oxidation process, as described by Fogg and Bhanot [39] in the 1980s, using stationary solid electrodes. In the work of Desimoni et al. [40], the authors used carbon glass electrode (GCE) modified with Nafion for the detection of Patent Blue (V) dye, by oxidation of the R–OH to R=O group. Under optimized conditions such as electrolyte (0.1 mol L^{-1} acetate buffer solution) and pH (5.0), an analytical curve was constructed in the interval from 9.5×10^{-8} to 9.9×10^{-7} mol L^{-1} using differential pulse voltammetry (DPV), found a detection limit of 7.6×10^{-8} mol L^{-1}. In another work, Nayak and Shetti [41] developed a glucose-modified carbon paste as sensor for erythrosine. The measurements were performed in phosphate buffer solution pH 11.2, since the sensor exhibited higher catalytic activity. The mechanism proposed by the authors is exemplified in **Figure 2**. Using square-wave voltammetry (SWV), an analytical curve was constructed between 1.0×10^{-7} and 1.0×10^{-4} mol L^{-1} with detection limit of 2.16×10^{-8} mol L^{-1}. The method was applied in human urine sample, with recoveries between 91.6 and 98.0% and relative standard deviation of 1.12%.

Zhang and collaborators [42] have proposed the determination of food dye Ponceau 4R and Allura Red using a multi-wall carbon nanotube-modified GCE. In this case, the dye signal occurred by the oxidation of the R–OH group for both analytes. By means of DPV measurements, linear relationships were found in pH 7.0 phosphate buffer 0.1 mol L^{-1} solution from 25 μg L^{-1} to 1.5 mg L^{-1} and 50 μg L^{-1} to 0.6 mg L^{-1} and detection limit of 15 and 25 μg L^{-1} for Ponceau 4R and Allura Red, respectively. The proposed method was applied in soft drinks samples with high accuraty and feasibly. In the work of Sierra-Rosales et al. [43], the authors used a GCE modified with multi-walled carbon nanotubes (MWCNTs) and 1,3-dioxolane as a dispersant agent for the determination of Tartrazine, Sunset Yellow, and Carmoisine. For the three dyes, the oxidation occurred by the loss of one $1e^-$ and one H^+ from oxidation of the R–OH group. Analytical curves were constructed by DPV in 0.1 mol L^{-1} phosphate buffer solution (pH 7.0) using 2 min of accumulation time. Linear regions were obtained in the interval from 1.0 to 7.0, 0.55 to 7.0, and 0.54 to 5.0 μmol L^{-1} and limits of detection of 0.22, 0.12, and 0.11 μmol L^{-1} for Tartrazine, Sunset Yellow and Carmoisine, respectively. Finally, the method was applied in soft drinks sample reaching recoveries from 87 to 109% for all the

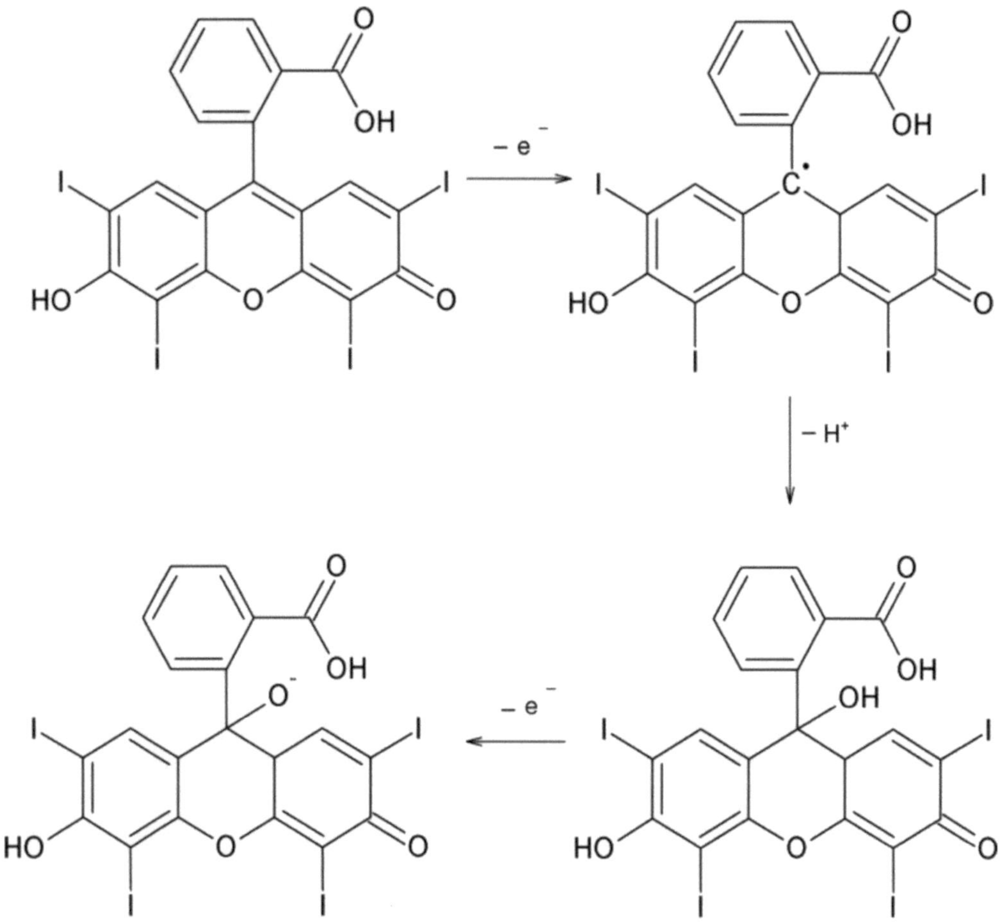

Figure 2. Oxidation mechanism of erythrosine on glucose-modified carbon paste sensor in basic phosphate buffer solution [41].

analyzed dyes. The method was compared with high-performance liquid chromatography (HPLC) method and showed great proximity between both methods.

3. Electroanalysis of hair dyes

The use of hair dyes with the purpose to change the look is a common practice for men and women. In the 1950s, only 7% of Americans usually had their hair colored. Nowadays, more than 75% of American women use hair dyes [44]. In the same way, in Japan, the hair dyeing process has grown from 14–41% in 1992–2001 to 60–85% in the last 20 years, mainly from high school women [45]. In Brazil, the IBOPE agency indicated that 26% of the population uses hair dyes, of which 85% are women and 15% are men [46].

The first records of hair dye were identified in Egypt, where the natural dye known as henna, extracted from the leaf of *Lawsonia inermis* plant, was found in the mummies' hair [47–49].

Some other natural dyes were also used in the past, such as chamomile. In addition to these types, in some Asian countries, it was found that the extracts from nutgall, logwood, and brazilwood were used to cover gray hair [50]. However, due to short variety of colors, the natural hair dyes lost space after the first development of synthetic hair [7, 51]. Since then, the great demand and use of these products have made the technological development and growth of this market seek new products with high fixation and low price.

Hair dyes are usually classified according to the fixation on hair. They can be assigned as direct dyes (semipermanent and temporary dyes) and oxidative dyes (permanent dyes) [50, 52]. Direct dyes are separated into semipermanent or temporary dyes. The first of these (semipermanent dyes) are low molecular weight substances derived from nitroanilines, nitrophenylenediamines, and nitroaminophenols [53]. The semipermanent dyes are responsible for 10% of the economy of hair dyes and are marketed with a mixture of 10–12 different dyes to achieve the desired shade, which can remain in the hair for up to 6 weeks. Due to the low molecular weight, it allows the dye to diffuse into the cortex region, developing Van der Waals and weak polar interactions [50]. However, the temporary dyes, that present high molecular mass, cannot permeate to the region of the cortex [54]. Thus, these dyes are deposited only due to Van der Waals interactions or simple adsorption on the hair, which justifies their short durability. The products marketed as temporary dyes, such as shampoo, sprays, and lotions, have in their composition a mixture of two to five different types of temporary dyes to acquire the desired shade [55].

The second class of dyes is the permanent dyes (or oxidative dyes). They can promote permanent fixation into the hair. The process involves opening of the cortex and interaction of the components of the dye with inner regions of the hair strand [7, 55]. Oxidative dyes are the most representative among hair dyes due to their versatility, easy application, and mainly high fixation, which is reflected commercially; they represent 80% of the economy in this sector, in the USA and EU [56]. The products marketed as permanent dyes are available in kits containing two components: the first component is a mixture containing precursor agents, which are aromatic amines such as *p*-phenylenediamine (PPD) and *p*-aminophenol, and coupling agents, which are electron donor substances such as resorcinol (RSN) and naphthol [53]. The second component consists of an oxidizing agent in alkaline media, since the hair bleaching process is more effective in basic solutions [50, 53]. When the two components are mixed, the oxidizing agent (i.e., hydrogen peroxide) promotes the oxidation of the precursor agent (usually PPD), forming intermediates such as *p*-quinoneimine (PDQ) [18]. After this process, the dye itself will begin to form from the reaction of the intermediate formed, PQD, with the coupling agent (usually RSN) [7]. The reactions involved in the dye formation process (**Figure 3**) occur within the hair, that is, upon penetration of the coupling, precursor and oxidizing agents, into the cortex.

However, besides the motivation for the development of a wide variety of shades for hair dyes, the human health risks of these substances have also been the subject of research. The main worry is about the additives that may contain azo compounds and other amine, nitro, and other derivatives, which may be hazardous for the human health [49, 57–59]. As an example, in 1975, AMES et al. [60] evaluated for the first time the mutagenicity of hair dye ingredients. In a recent work, Hudari et al. [18] have shown that when PPD is used as a precursor agent, besides the formation of dyes, there is also the formation of the trimer called Bandrowski's Base (**Figure 3**), which is associated with several allergic reactions and possible carcinogenic properties [61–63].

For these reasons, several electroanalytical methods have been proposed for the identification and/or determination of hair dyes and their ingredients. In the early 1960s, Olson and co-authors [64] studied the electrochemical behavior of *p*-nitroaniline on the surface of a carbon paste electrode. They established that reduction of the $-NO_2$ groups involves $6e^-$ and can form amine products [64, 65]. In the work of Tong et al. [66], the authors used a rotating disk electrode to study the PPD reactions at the electrode. For quantitative purposes, Lawrence and coauthors [67] use a GCE for determination of PPD by cyclic voltammetry (CV) and SWV. PPD presented a reversible behavior regarding the oxidation of the $-NH_2$ groups and subsequent reduction of the groups =NH, with the participation of $2e^-$ and $2H^+$ in oxidation and reduction processes. Analytical curves were constructed using the CV and SWV, where linear relationships were found between 2–200 μmol L^{-1} and 2–20 μmol L^{-1} and limits of detection was 1.2 and 0.6 μmol L^{-1}.

Aiming for better levels of detection and selectivity, the search for new electrode materials and the modification of the surface of the electrodes became preponderant for the development of electroanalytical methods for the determination of hair dyes. As an example of new electrode materials, Oliveira and Zanoni [68] have used a self-organized Ti/TiO_2 nanotubular array electrode to monitor the reduction of azo group to hydrazo group in the hair dye basic brown 17 after a process involving $2H^+$ and $2e^-$. After optimization of SWV technique, an analytical curve was constructed in the interval from 1×10^{-7} to 7×10^{-6}mol L^{-1}. The method exhibited a limit detection of 2.7×10^{-9} mol L^{-1}. As an example of the use of modifiers, Hudari et al. [18] proposed a composite carbon nanotube/chitosan for the modification of the surface of a GCE

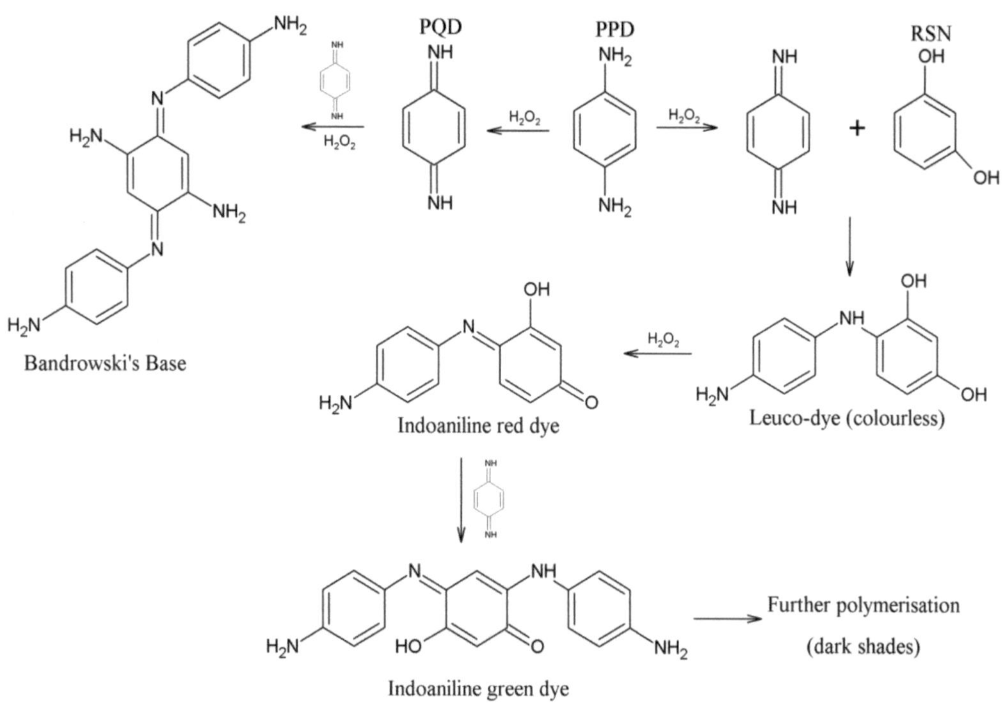

Figure 3. Reactions involved in the formation of permanent dyes using PPD, RSN, and H_2O_2 as the precursor, coupling, and oxidizing agents, respectively [49].

for the determination of the precursor and coupling agents *p*-phenylenediamine (PPD) and resorcinol (RSN), respectively, in commercial permanent dye sample. The determination and subsequent quantification of the analytes was done through the oxidation of the $-NH_2$ to $=NH$ groups (involving $2H^+$ and $2e^-$) and the –OH to =O groups (involving $2H^+$ and $2e^-$) present in the PPD and RSN molecules, respectively. Using linear sweep voltammetry (LSV) and measurements in 0.1 mol L^{-1} ammonia buffer (pH 8.0) solution, a linear relationship was found in the range of 0.55–21.2 mg mL^{-1} for both analytes with detection limits of 0.79 and 0.58 mg mL^{-1} for PPD and RSN, respectively. Finally, the method was applied in a sample of commercial hair dye and compared with the UV–vis spectrophotometric method, which, with great agreement between the two methods. In the work of Zhao and Hao [69], the authors developed a molecular imprinting electrochemical sensor for the determination of 2,6-Diaminopyridine. Similar to PPD, the 2,6-Diaminopyridine exhibits a reversible behavior due to the oxidation of the $-NH_2$ group and subsequent reduction of the =NH group. A wide linear range was found between 0.0500 and 35.0 mg kg^{-1} with limit of detection of 0.0275 mg kg^{-1}. The method was applied in samples of commercial dyes with a recovery from 98.40 to 103.8% and relative standard deviation of less than 1.51%. Corrêa et al. [70] used a composite electrode to preconcentrate carboxyl-functionalized magnetic nanoparticles for the determination and quantification of the Basic Brown 16. In this case, the determination of dye was made by the signal referring to the oxidation of the –OH group, which allowed a linear relationship between 1.00×10^{-7} and 1.00×10^{-6} mol L^{-1} with limits of detection and quantification of 1.01×10^{-8} and 2.37×10^{-8} mol L^{-1}, respectively. Finally, the method was applied with great success for the determination of hair dye in wastewater and also in samples of dye removed from dyed hair strands.

In recent years, miniaturized systems, such as printed electrodes, have also deserved attention. Disposable electrodes have been widely used as a viable alternative for rapid measurements of dye in low concentrations [71–73]. As an example, Hudari and coauthors [16] proposed a method for the determination of hair dye Basic Blue 41 using screen-printed carbon electrodes modified with graphene. The well-defined peak is attributed to the reduction of the azo group. After multivariate optimization of SWV instrumental parameters, such as frequency (54.8 Hz), pulse amplitude (43.7 mV), and step potential (6 mV), an analytical curve was constructed in the range of 3.00×10^{-8} to 2.01×10^{-6} mol L^{-1} with limits of detection and quantification of 5.00×10^{-9} and 1.70×10^{-8} mol L^{-1}, respectively. The sensor was successfully applied in wastewater samples and validated by comparison with HPLC-DAD method with good accuracy.

4. Electroanalysis of textile dyes

Textile dyes are a class of colored substances used to impart permanent color to textile fibers. The most used class of dye is the azo dyes that present low cost and great diversity of colors and other promissory characteristics [74, 75]. The dye can be fixed to the fiber by several mechanisms, mainly due ionic Van der Waals and hydrogen interactions, but also due covalent bond [76].

The ionic bonds are frequently found in the dyeing of wool, silk, and polyamide. It came from the interactions between oppositely charged ions of the dye bearing charged groups and in the protonated groups in the fibers [76]. The Van der Waals has been found in dyeing of wool and polyester. It results from an approach between the π orbitals of the fiber and dye molecule [76]. Hydrogen interactions have been found in the dyeing of wool, silk, and synthetic fibers, such as ethyl cellulose. It has resulted from the interaction between hydrogen atoms covalently bonded in the dye and free electron pairs of donor atoms in the center of the fiber [76]. Covalent bonds have been used in cotton fiber dye. They are formed between reactive groups (electrophilic groups) of the dye molecule and nucleophilic groups on the fiber [76, 77].

On the other hand, the textile dyes can also be classified based on the chemical structure (azo, antraquinone, etc.) or based on the method used to transfer the dye to the fiber. The main classes of textile dyes are reactive, direct, azoic, acid, vat dyestuffs, sulfur, disperse premetalized, and bleaching [76]. The main characteristics of this class are show below.

Reactive dyes bear in their structure an electrophilic group (reactive) that can form covalent bond with hydroxyl groups, amino groups, and thiol groups present in cellulose fibers, wool and polyamides for instance. In most cases, the reactive dyes bear an azo group or anthraquinone group, like chromophores, ethyl sulfonyl sulfate, and chlorotriazine, as reactive center. They are highly soluble in water and commonly used in dyeing of cellulose like cotton or flax, but also wool is dyeable with reactive dyes. Among the reactive dyes, the most used are the azo dyes because the azo group (−N=N−) confers to these dyes resistance to light, acids, bases, and oxygen. However, these desired proprieties makes them hazardous for the environment even at low concentration.

Direct dyes are anionic dyes soluble in water and can be used to dye cellulose fibers (rayon, silk, and wool) by Van der Waals interactions. They can be used for cellulosic fibers, normally applied from an aqueous dyebath containing an electrolyte, either sodium chloride (NaCl) or sodium sulfate (Na_2SO_4). They mostly show chromophore azo groups (diazo, triazo).

Azoic dyes are products insoluble in water and cannot be applied directly on fibers as dyes. They are produced within the fibers itself. Acid dyes are the major class of anionic dyes that present sulfonic groups, for instance, to increase the solubility in water and to promote a dye bearing negative charge to the dye molecules. The textile acid dyes are effective for protein fibers such as silk, wool, nylon, and modified acrylics. These dyes are characterized by substances with a chemical structure with the presence of azo, anthraquinone, triarylmethane, azine, xanthine, ketonimine, nitro, and nitrous groups, which provide a wide color range and degree of fixation.

Vat dyestuffs are mainly based on indigos, toringoides, and anthraquinone dyes. They are slightly soluble in water; in the dying process, they are reduced with dithionite in alkaline solution, transforming into a soluble compound (leuco form), and they are subsequently oxidized by air or another reagent, such as hydrogen peroxide, regenerating the original form of the dye on the fiber. In this type of dye, the carbonyl group can be present in an ethylenic group or alicyclic subunits. The main application of this type of dye has been the cotton dye.

Sulfur dyes are characterized by macromolecular compounds with polysulfide bridges ($-S_n-$), which are highly insoluble in water. First, they are applied after prereduction in sodium dithionite, which gives them the soluble form; these are subsequently reoxidized onto the fiber by contact with air. These compounds have been used mainly in dyeing of cellulosic fibers, imparting colors such as black, olive green, marine blue, and brown.

Disperse dyes are scarcely water-soluble dyes, originally used for dyeing synthetic fibers, and usually applied from fine aqueous suspensions and the presence of surfactants. They are used in hydrophobic fibers, such as cellulose acetate, nylon, polyester, and polyacrylonitrile.

Premetalized dyes are characterized by the presence of a hydroxyl or carboxyl group in the ortho position of azo chromophore, allowing the formation of complexes with metallic ions. They are useful mainly for dying protein fibers and polyamide.

Bleaching dyes are a class of compounds that have been used to decrease the color of natural textile fibers. The process involves the dyeing of fiber with chemical bleaches or white dyes, also known as optical brighteners or fluorescent brighteners. These dyes present carboxylic groups, azomethine (–N=CH–) or ethylenic (–CH=CH–) groups allied to benzene, naphthalene, pyrene and aromatic rings.

During the dyeing process, usually, there is a loss of around 10–50% of the dye to the environment [77, 78]. Considering that some dyes are highly toxic and mutagenic, and can also disturb the light penetration in surface water and therefore the photosynthetic process [79], their discharge deserves attention. The literature reports that azo dyes represent 60–70% of all organic dyes produced in the world [74], and they can be easily reduced to amine groups. So, the development of efficient and low-cost methods for the determination of this kind of dye is very important; electroanalytical method has been shown to be a successful alternative to this end.

The electrochemical reduction of two reactive dyes, Procion Red HE-3B 9 (RR120) and Procion Green HE-4BD (RG19), has been reported by cyclic voltammetry, differential pulse and DC polarography, chronoamperometry, and controlled potential electrolysis at mercury electrodes [80]. These dyes can be present in dyebath and wastewaters, considering that in the dye-fiber reaction the efficiency varies from 50 to 90%. Procion Red HE-3B (RR120) and Procion Green HE-4BD (RG19) are dyes with bis-azo groups that are reduced to bis-hydrazo derivative after transfer of four electrons generating the hydrazo derivative. Linear correlations in the range 1.0×10^{-7} mol L^{-1} to 1.0×10^{-5} mol L^{-1} were obtained between peak current and dye concentration for the peaks due to the reduction of both bis-azo and bis-monochlorotriazine group from the. The method allowed a limit of detections of 0.1 μmol L^{-1} for both dyes [80].

The electrochemical oxidation of Reactive Black 5 (RB5), a vinyl sulphone azo dye, has been also proposed by Radi et al. [81] at glassy carbon electrode (GCE) in phosphate buffer solutions in the pH range 2.85–11.79, employing cyclic voltammetry (CV) and differential pulse voltammetry (DPV). Under alkaline conditions, these dyes react with the hydroxyl groups of cellulose, by nucleophilic substitution or forming a covalent bond. RB5 presents a well-defined oxidation peak at 0.614 V *vs* Ag|AgCl using DPV in phosphate buffer at pH 4.20 that can be used for this determination from 6.0×10^{-7} to 15×10^{-6} mol L^{-1} in phosphate buffer (0.2 mol L^{-1}, pH 4.20). The limit of detection and quantification were 4.0×10^{-7} and 1.1×10^{-6} mol L^{-1}, respectively.

Reactive Red 231 has been also determined by glassy carbon electrode by cyclic and differential pulse voltammetry on a glassy carbon electrode in phosphate buffer [82]. The dye oxidation shows two peaks, attributed to the following process shown in **Scheme 1**. Using anodic differential pulse voltammetry in pH 3.77, analytical curves were constructed from 0.25 mmol L^{-1} to 200 mmol L^{-1}. The method offers a reproducibility of RSD 4.7% and a detection limit of 0.20 mmol L^{-1}. The method has been applied to determine the dye in environmental water samples, with recoveries from 85 to 97%.

Turquoise blue 15 (AT15) is a reactive dye widely used in the textile industry to color natural fibers and has been determined in aqueous solution at mercury and glassy carbon electrode [83]. Using linear cathodic stripping voltammetry in acidic medium, the copper phthalocyanine is reduced in only one step, but it exhibits two reduction waves in an alkaline medium at glassy carbon electrode. The reduction of AT15 at mercury electrode (HMDE) is similar to the reduction at glassy carbon electrode, shown three cathodic peaks. These peaks can be attributed to ligand reduction. In addition, an extra peak could be observed at less negative potential, with preaccumulation, attributed to the reduction of the metal center. So, the results suggest a adsorption of the dye on mercury electrode stabilized by coordination bond metal/phthalocyamine ring first, and next the reduction of Cu(II) and pyrrole ring. Using the best experimental conditions, linear analytical curves were obtained for the first reduction step from voltammograms recorded at accumulation time of 180 s from 1.00×10^{-8} mol L^{-1} to 1.00×10^{-7} mol L^{-1}. The repeatability of the method shows a relative standard deviation from 1.36 to 2.05%. Detection limits were estimated from 8.34×10^{-9} to 1.50×10^{-7} mol L^{-1}, and the method was applied with success in tap water and in textile plant effluent sample [83].

The determination of Black 5 and Red E was determined by quartz crystal resonator sensor coated with mesoporous carbon cryogel [84]. The determination of the dyes is based on the vibrational motion of the plate in a resonant frequency, where the frequency is sensitive to mass loading on the electrode. Analytical curves were obtained from 25 to 100 ppm. The

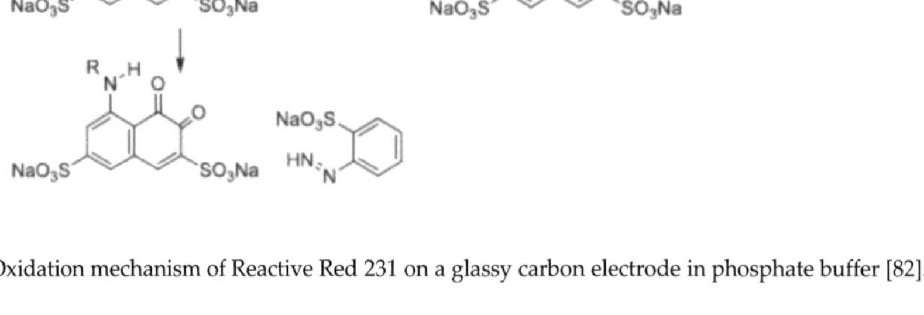

Scheme 1. Oxidation mechanism of Reactive Red 231 on a glassy carbon electrode in phosphate buffer [82].

method is selective since the method is based on the size of each reactive dye and the micropores of the activated carbon resonator.

The analysis of vat dyes Indanthrene Olive Green B dye (VG3), which present an anthraquinonoid group and a ketonic group, has been also reported by differential pulse voltammetry in alkaline solution using glassy carbon electrode [3]. A typical linear scan voltammogram obtained for VG3 dye presents three reduction cathodic processes at −0.57, −0.67, and −0.99 V. A good linear correlation from 1.0×10^{-4} to 7.0×10^{-4} mol L^{-1} was obtained in sodium hydroxide 0.1 mol L^{-1} pre-accumulated during 30 s at 0 V on glassy carbon electrode. The detection limit was 5×10^{-5} mol L^{-1}.

Santos et al. [85] has also reported the analysis of the disperse dye Disperse Red 13 (DR13) at glassy carbon electrodes (GCEs) modified with polyglutamic acid (PGA). This dye bears reductive nitro and azo group and presented a well-defined peak at −0.66V when reduced in a mixture of DMF/BR (pH 4, 1:1 v/v) and Bu_4NBF_4/DMF. At glassy carbon electrode modified with polyglutamic acid, the cathodic peak of DR13 shifts at least 200 mV to a less negative potential, and the peak intensity is 2.5 times higher compared to bare electrode. A linear calibration curve was obtained in DMF/BR buffer (pH 4) from 2.5×10^{-7} to 3.0×10^{-6} mol L^{-1}. The limit of detection was 1.5×10^{-8} mol L^{-1} and the repeatability presented an RSD of 4.29%. The method was applied in river water sample and dyebath wastewater.

The poly-L-lysine-modified glassy carbon electrode was also applied in determination of Cibacron Blue F3GA [17]. The dye presents an anthraquinone group as chromophore and amine group that is oxidized at 0.75 V. A linear calibration curve was obtained from 1.0×10^{-6} to 1.0×10^{-5} mol L^{-1} in BR buffer pH 2.0 after a preconcentration off-line by 10 min. The method reached a limit of detection of 4.5×10^{-8} mol L^{-1}, and it was applied in tap water and raw water sample.

The simultaneous determination of Orange G (Or G) and Orange II (Or II) in industrial wastewater has been proposed by using carbon paste electrode containing Fe_2O_3 nanomaterials/oxygen functionalized multiwalled carbon nanotubes/triton X-100 modified (Fe_2O_3/MWCNTs-COOH/OP/CPE) [86]. The optimized conditions for dyes analyses were 10 μL of MWCNTs-COOH/OP dispersion on the MWCNTs-COOH/CPE surface, accumulation time of 4 minutes, pH 7 of supporting electrolyte 0.1 mol L^{-1}. A liner calibration curve was observed by DPV method from 0.1–20.0 μmol L^{-1} to 0.2–50.0 μmol L^{-1} and detection limits of 0.05 and 0.1 μmol L^{-1} for Or G and Or II, respectively. The developed sensor was applied to the simultaneous detection of Or G and Or II in different industrial wastewater samples. The RSD was lower than 5%, the recoveries for Or G and Or II in these samples ranged from 96.8 to 105.1% and from 97.4 to 103.8%, respectively.

The determination of direct orange 8 dye in a flow stream was developed by using a wall-jet electrode system and square-wave stripping method [87]. The DO8 shows an electrochemically oxidizable phenolic group at −0.65V. After optimized experimental conditions, and using stripping voltammetric method, a linear calibration curve was obtained from 1.0×10^{-4} to 1.5 mg mL^{-1}. The method shows a reproducibility of 2.0 % and recovery of 98%.

Abrasive stripping voltammetry (AbrSV), based on a mechanical transfer of material onto the surface of a solid electrode and the subsequent voltammetric measurement of the electrochemical stripping process, has been proposed by Chen et al. [88]. They analyzed six different solid compounds of widely different natures in room temperature ionic liquids (RTILs), among them the Prussian blue (PB) and indigo dye. In PB analysis, a reversible oxidation and reduction peak centered at around 0.6 V were observed. The authors indicates that the reduction and oxidation reactions originated from a surface redox species: $Fe_4^{III}[Fe^{II}(CN)_6]_3 + M^+ + e^- \rightarrow MFe^{II}Fe_3^{III}[Fe^{II}(CN)_6]_3$, where M^+ could be a cation, such as K^+, Li^+, or H^+. In addition, indigo electroanalysis shows a large reduction peak at −1.19 V, corresponding to indigo reduction: $H_2IN \rightarrow H_2IN_2^-$.

Hydrodynamic electrode system based on a vibrating probe (250 Hz, 200 lm lateral amplitude) has been applied to determine indigo in a complex plant of indigo sample, with high levels of organic and inorganic impurities after the reduction by glucose in aqueous 0.2 mol L^{-1} NaOH [89]. The soluble leucoindigo is determined by oxidation response at the vibrating electrode. In alkaline media, indigo is reduced in the presence of glucose to give leucoindigo that does not interfere in the electrode process. The best analytical signal was obtained using high scan rate, low vibration amplitude (250 Hz), and high temperature (75°C). The method was applied in 25 different samples of plant-derived indigo.

Finally, the determination of alizarin red S (AR), an anthraquinone dyes used in textile industry, has been proposed by flow-through potentiometric sensor [90]. The sensor is based in the use of a sensitive poly (vinyl chloride) membrane and the use of aliquate 336, Mg^{II}phthalocyanine (MgPc), Cu^{II}phthalocyanine (CuPc), and Fe^{II}phthalocyanine (FePc) plasticized poly (vinyl chloride) membrane. The sensor allowed detection limits of 5.9×10^{-7}, 1.9×10^{-6}, 2.3×10^{-6}, and 1.9×10^{-6} mol L^{-1} for aliquate, MgPc, CuPc, and FePc membrane-based sensor, respectively, and accuracy higher than 99.4%. The method shows a linear calibration curve from 0.1–1.8 to 1.0−40 μg mL^{-1} and a detection limit of 0.08 and 0.5 μg mL^{-1} for 10^{-4} and 10^{-3} mol L^{-1} AR as a carrier, respectively.

5. Electroanalysis of marker dyes

The marker dye are used mainly in the groundwater protection and in the prevention of tax frauds, such as identification of fuel with high sulfur concentration [91, 92]. These dyes are cataloged based on their characteristics and following a standard that depends on the type of dye and degree of coloring. An example of dyes commonly used as marker is the Solvent red 24 and Solvent blue 35, structures of which are shown in **Table 2**. The "Solvent red" and "Solvent orange" dyes usually bear azo group in their structure unlike the "solvent green" and "solvent blue" that bears anthraquinone groups. These dyes are used in fuel coloring due their higher solubility in organic solvents, nonpolar and slightly polar. They are commonly used as marker solvent in petroleum derivatives, such as waxes, lubricants, plastics, and other nonpolar materials. The marker dye usually has the chemical structure protected and only the manufacturer can detect it by specific methods. Neverthelees, some marker dyes are known in the literature and are commonly added to fuels with the purpose of control specific types of fuels, comprove the authentic, discourage robberies and adulterations, as well as control the distribution and use of some specifics fuels [93, 94]. In Brazil, the adulteration of fuel has been a common problem, and the marked dyes are applied to distinguish the hydrated ethanol from

http://dx.doi.org/10.5772/67945

the dehydrated ethanol. In agreement with Brazilian laws, a solvent dye with orange coloration is prescribed to be added to dehydrated ethanol as a control. The aviation fuels are also marked as a guarantee to prevent fraud or other dilutions [95, 96].

In many countries, particularly those in the European Community, fuel distributors use marker dyes to diesel, "the red diesel," since it is significantly cheaper than heavier diesel fuels with a higher sulfur concentration that can promote damages to both the environment and the engine. In view of such problems, enforcement agencies have adopted the addition of marker dyes as a method of prevention as a way to curb the adulteration of these types of fuels. Among these dyes, red dyes, which are present in azo groups, such as Red Solvent 19, Red Solvent 24, and Red Solvent 26, are prominent. The aviation kerosene, as well as heating oils and so on, has also been marker with dyes, whose purpose is to prevent their mixing with poor-quality products. Nowadays, all the countries of the European Union are being obliged to adopt a marker dye; the most common is the "Solvent Yellow" 124. This dye can easily detect in adulterated fuels by dilution

Anthraquinone group	**Azo group**
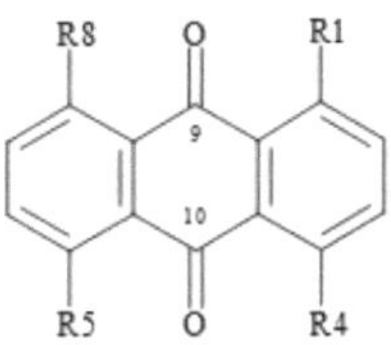	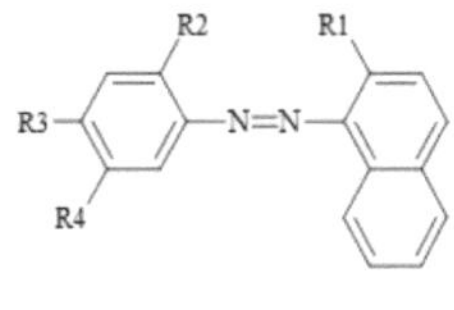

Dye	R₁	R₂	R₃	R₄	R₅	R₈
	Ring substituent					
Anthraquinone						
Solvent Blue 14		H	H		H	H
Solvent Blue 36	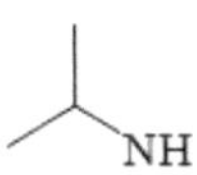	H	H	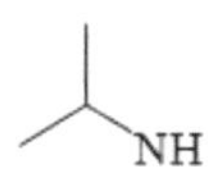	H	H
Solvent Blue 35		H	H		H	H
Solvent Green 3	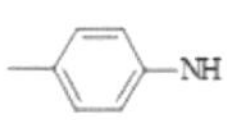	H	H	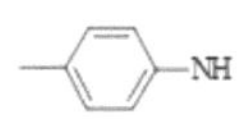	H	H
Azo						
Solvent Red 24	OH	H	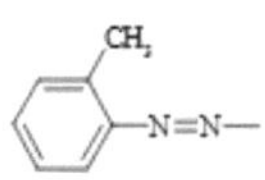	H	–	–
Solvent Red 26	OH	CH_3	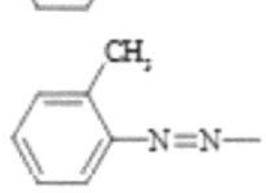	CH_3	–	–
Solvent Red 164	OH	H	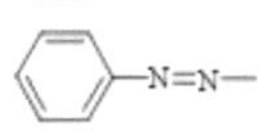	H	–	–
Solvent Yellow 14	OH	H	H	H	–	–

Table 2. Molecular structures of marker dyes.

with solvents or in poor-quality products at very low levels [97]. In the USA, the Environmental Protection Agency (EPA) designates the use of a red dye, the "Red Solvent 26," to identify and label high sulfur fuels [98]. In Brazil, in recent years, a fuel alcohol marking system has been adopted [99], whose purpose is to ensure its quality, as well as to free it from adulteration fraud by addition of water. The marker dyes, such as Quinizarine (QNZ), solvent blue 14 (S-14), Solvent Blue 14, Solvente Blue 35 (SB-35), Solvent Orange 7 (SO-7), and Red Solvent 24 (SR-24) can be legally added for the purpose of meeting the requirements of the ANP, or illegally, for circumventing inspection agencies.

Thus, analytical methods, able to determine these marker dyes and help the protection of quality of fuel, are highly demanded. Some electrochemical methods are proposed in the literature with this purpose and they are resumed below.

The determination of solvent blue 14 (SB−14), 1,4-bis(pentylamino)anthraquinone, in fuel samples has been studied by square-wave voltammetry (SWV) at glassy carbon electrode in a mixture of aqueous solution and N,N-dimethylformamide [100]. SB-14 dye presented a well-defined cathodic peak at around −0.41 V *vs* Ag|AgCl attributed to the reduction of the central quinone group in the dye molecule. The best conditions for SB-14 determination were obtained in a mixture DMF:BR buffer pH 2.0, using square-wave voltammetry operating at *f*: 60 Hz, ΔE_s: 6 mV and E_{sw}: 50 mV. Under optimized conditions, there is a linear response from 1.0×10^{-6} to 6.0×10^{-6} mol L^{-1} and a detection limit of 2.90×10^{-7} mol L^{-1}. The method was applied to dye SB-14 determination in kerosene, after cleanup with a self-packed SPE column (SiO_2) and alcohol samples with average recovery from 93.00 to 98.10%.

A similar method has been also developed for determining solvent Orange 7 (SO-7), an azo dye in fuel ethanol samples [4]. The optimal conditions adopted a mixture of N,N-dimethyl-formamide (DMF) and Britton–Robinson buffer pH 7.0 (1:1, v/v). In this medium, a well-defined anodic peak at +0.70 V *vs* Ag|AgCl was observed. Using optimized parameters based on the SWV technique, it was possible to construct a linear relationship from 4.0×10^{-6} to 18.0×10^{-6} mol L^{-1}. The proposed method was successfully applied to the direct quantification of the SO-7 dye in fuel ethanol samples, with recovery between 97.2 and 106%.

The screen-printed carbon electrode in Britton–Robinson buffer with N,N-dimethylformamide (7:3, v/v) in presence of 5.50×10^{-4}mol L^{-1} of dioctyl sulfosuccinate sodium (DSS) has been reported to analyze solvent blue 14 (SB-14), 1,4-bis(pentylamino)anthraquinone [101]. The SB-14 showed a cathodic peak at −0.40 V due to reduction of the central quinone group to hydroquinone derived after a two-electron process. Using the optimal conditions, $f = 60$Hz, $\Delta E_s = 4$mV and $E_{sw} = 50$mV, and pH 3.0, an improved interaction between the dye and the anionic surfactant by electrostatic interaction was obtained in a mixture of B–R buffer (pH 3.0)/DMF (7:3, v/v) + 1.40×10^{-3}mol L^{-1} DSS surfactant. A calibration curve was constructed from 2.00×10^{-7} to 2.00×10^{-6}mol L^{-1} with limit of detection and quantification of 9.30×10^{-8} and 3.0×10^{-7} mol L^{-1}, respectively. The developed method was applied for the quantification of SB-14 dye in alcohol samples without any pretreatment and in kerosene (after rapid cleanup procedure based in solid-phase extraction cartridge) with recoveries of 82.00–99.00%, respectively. The values are in good accordance with other reference methods based on spectrophotometric analysis.

http://dx.doi.org/10.5772/67945

The 1,10-dihydroxyanthraquinone, quinizarine (QNZ), has been determined by square-wave voltammetric technique in a mixture of Britton-Robinson buffer 0.08 mol L^{-1} with 30% of acetonitrile [102]. The QNZ was oxidized at glassy carbon electrode in and the well-defined peak at +0.45 V *vs* Ag|AgCl. This peak can be attributed to the oxidation of the phenolic hydroxyl group present in the dye marker to its quinone derivative after two electrons' transfer. An analytical curve was constructed for QNZ concentrations ranging from 2.0×10^{-6} to 1.4×10^{-5} mol L^{-1}. The method was successfully applied for determining QNZ in gasoline and diesel oil, the recoveries were 94.20 and 90.80%, respectively. For diesel oil sample, the QNZ was first extracted by liquid-liquid extraction using ACN: BR, 7:3 v/v and next was performed a solid-phase extraction cartridge of C18 adopting.

Quinizarine has been also determined by square-wave voltammetric technique using screen-printed carbon electrode (SPCE) in the presence of cetyltrimethylammonium bromide (CTAB) [103]. In the presence of 7.50×10^{-4} mol L^{-1} CTAB in a mixture of BR buffer (pH 7.0) and DMF (7:3, v/v), quinizarine (QNZ) shows a cathodic peak at −0.68 V, without CTAB the peak was observed at −0.78 V. The peak in the presence of CTAB was three times higher than the one obtained in its absence. These peaks are both attributed to the electron-transfer involving the reduction of central quinone group to form hydroquinone derived after a two-electron process. The electroanalytical method presented a linear response from 5.00×10^{-7} to 6.00×10^{-6} mol L^{-1} and a detection limit of 2.70×10^{-7} mol L^{-1}. The methodology was applied in the dye marker quantification in diesel oil and kerosene samples with recovery values ranging between 84.0 and 98.7%.

6. Conclusion

The voltammetric methods have shown to be a powerful tool for dye determination considering the great diversity of dye and different utilities. Different voltammetric techniques have been proposed, such as differential pulse, square-wave stripping, abrasive stripping hydrodynamic, polarography, chronoamperometry, and others. These techniques minimizing the effect of capacitive current (techniques of differential pulse voltammetry and square-wave voltammetry) and possibility of preconcentration of the analyte at the electrode surface promoted analysis applicable at lower detection levels. Furthermore, different electrodes have been proposed, such as dropping mercury electrode, glassy carbon, screen-printed electrode, Ti/TiO_2 nanotubular array electrodes, and others. The use of solid stationary electrodes improved the applicability of the method, since they can be easily modified and offer the possibility to increase the sensitivity and selectivity of the selected dye. The main modifiers found in literature are based on the use of metal nanoparticles, polyaminoacids, carbon nanotubes, graphene, molecularly imprinted polymers, and others that offers higher sensitivity and/or selectivity. So, the voltammetric methods allow the detection of dyes in different matrix in concentration level until 10^{-9} mol L^{-1}, similar to other instrumental techniques such as chromatography. The main advantages of voltammetric methods are that they are low cost, are environmental friendly and quick, and, in many cases, do not require long steps of cleanup sample.

Author details

Felipe Fantinato Hudari[1], Michelle Fernanda Brugnera[2]* and Maria Valnice Boldrin Zanoni[1]

*Address all correspondence to: michellebrugnera@gmail.com

1 Department of Analytical Chemistry, Institute of Chemistry of Araraquara, UNESP, Araraquara, SP, Brazil

2 Department of Chemistry, Earth and Exact Sciences Institute, Federal University of Mato Grosso, Cuiabá, Brazil

References

[1] P. Gregory. Dyes for polyacrylonitrile. In: The chemisty and application of dyes. Boston, MA: Springer, US; 1990, pp. 165–201. doi:10.1007/978-1-4684-7715-3_5.

[2] T.M. Lizier, T.B. Zanoni, D.P. Oliveira, M.V.B. Zanoni. Electrochemical reduction as a powerful tool to highlight the possible formation of by-products more toxic than Sudan III dye. Int. J. Electrochem. Sci. 7 (2012) 7784–7796. http://www.electrochemsci.org/list12.htm#issue10

[3] M.V.B. Zanoni, W.R. Sousa, J.P. de Lima, P.A. Carneiro, A.G. Fogg. Application of voltammetric technique to the analysis of indanthrene dye in alkaline solution. Dye. Pigment. 68 (2006) 19–25. doi:10.1016/j.dyepig.2004.12.008.

[4] D.C. Romanini, M.A.G. Trindade, M.V.B. Zanoni. A simple electroanalytical method for the analysis of the dye solvent orange 7 in fuel ethanol. Fuel. 88 (2009) 105–109. doi:10.1016/j.fuel.2008.07.022.

[5] C. O'Neill, F.R. Hawkes, D.L. Hawkes, N.D. Lourenço, H.M. Pinheiro, W. Delée. Colour in textile effluents—sources, measurement, discharge consents and simulation: a review. J. Chem. Technol. Biotechnol. 74 (1999) 1009–1018. doi:10.1002/(SICI)1097-4660(199911)74:11<1009::AID-JCTB153>3.0.CO;2-N.

[6] L. Pereira, M. Alves. Dyes—environmental impact and remediation. In Environmental protection strategies for sustainable development. Dordrecht: Springer; 2012, pp. 111–162. doi:10.1007/978-94-007-1591-2_4.

[7] P. Ghosh, A.K. Sinha. Hair colors: classification, chemistry and a review of chromatographic and electrophoretic methods for analysis. Anal. Lett. 41 (2008) 2291–2321.

[8] E. Heidarizadi, R. Tabaraki. Simultaneous spectrophotometric determination of synthetic dyes in food samples after cloud point extraction using multiple response optimizations. Talanta. 148 (2016) 237–246. doi:10.1016/j.talanta.2015.10.075.

[9] A.H. El-Sheikh, Y.S. Al-Degs. Spectrophotometric determination of food dyes in soft drinks by second order multivariate calibration of the absorbance spectra-pH data matrices. Dye. Pigment. 97 (2013) 330–339. doi:10.1016/j.dyepig.2013.01.007.

[10] S. Sahin, C. Demir, S. Güçer. Simultaneous UV–vis spectrophotometric determination of disperse dyes in textile wastewater by partial least squares and principal component regression. Dye. Pigment. 73 (2007) 368–376. doi:10.1016/j.dyepig.2006.01.045.

[11] Y.S. Al-Degs. Determination of three dyes in commercial soft drinks using HLA/GO and liquid chromatography. Food Chem. 117 (2009) 485–490. doi:10.1016/j.foodchem.2009.04.097.

[12] M.J. Culzoni, A.V. Schenone, N.E. Llamas, M. Garrido, M.S. Di Nezio, B.S. Fernández Band, H. C. Goicoechea. Fast chromatographic method for the determination of dyes in beverages by using high performance liquid chromatography—diode array detection data and second order algorithms. J. Chromatogr. A. 1216 (2009) 7063–7070. doi:10.1016/j.chroma.2009.08.077.

[13] S. Dong, L. Chi, S. Zhang, P. He, Q. Wang, Y. Fang. Simultaneous determination of phenylenediamine isomers and dihydroxybenzene isomers in hair dyes by capillary zone electrophoresis coupled with amperometric detection. Anal. Bioanal. Chem. 391 (2008) 653–659.

[14] M. Narita, K. Murakami, J.-M. Kauffmann. Determination of dye precursors in hair coloring products by liquid chromatography with electrochemical detection. Anal. Chim. Acta. 588 (2007) 316–320.

[15] D.A. Skoog, F.J. Holler, T.A. Nieman. Princípios de Análise Instrumental. 5th ed. São Paulo: Artmed, 2013. 1056 p. ISBN: 9788577804603.

[16] F.F. Hudari, S.L. Costa Ferreira, M.V.B. Zanoni. Multi-responses methodology applied in the electroanalytical determination of hair dye by using printed carbon electrode modified with graphene. Electroanalysis. 28 (2016) 1085–1092. doi:10.1002/elan.201501043.

[17] E.R.C.C. Viana, F.C. Pereira, M.V.B. Zanoni. Electrochemical reduction and determination of Cibacron Blue F3GA at poly-L-lysine modified glassy carbon electrode. Dye. Pigment. 71 (2006) 145–152. doi:10.1016/j.dyepig.2005.04.020.

[18] F.F. Hudari, L.C. de Almeida, B.F. da Silva, M.V.B. Zanoni. Voltammetric sensor for simultaneous determination of p-phenylenediamine and resorcinol in permanent hair dyeing and tap water by composite carbon nanotubes/chitosan modified electrode. Microchem. J. 116 (2014) 261–268. doi:10.1016/j.microc.2014.05.007.

[19] M. Carocho, P. Morales, I.C.F.R. Ferreira. Natural food additives: quo vadis? Trends Food Sci. Technol. 45 (2015) 284–295. doi:10.1016/j.tifs.2015.06.007.

[20] Marcelo Alexandre Prado, Helena Teixeira Godoy. Corantes artificiais em alimentos. Alim. E Nutr. Araraquara. 14 (2003) 237–250.

[21] H.-D. Belitz, W. Grosch, P. Schieberle. Food additives. In: Food chemistry (4th revise). Berlin and Heidelberg: Springer; 2009, pp. 429–466. doi:10.1007/978-3-540-69934-7_9.

[22] Food and Drug Administration-Color Additives: Questions and Answers for Consumers (n.d.) Retrieved 13 December 2016, from http://www.fda.gov/Food/IngredientsPackagingLabeling/FoodAdditivesIngredients/ucm488219.htm

[23] A. Downham, P. Collins. Colouring our foods in the last and next millennium. Int. J. Food Sci. Technol. 35 (2000) 5–22. doi:10.1046/j.1365-2621.2000.00373.x.

[24] P. Gregory. Classification of dyes by chemical structure. In: The chemistry and application of Dyes (1st ed.). Boston, MA: Springer US; 1990, pp. 17–47. doi:10.1007/978-1-4684-7715-3_2.

[25] N.P. Boley, N.G. Bunton, N.T. Crosby, A.E. Johnson, P. Roper, L. Somers. Determination of synthetic colours in foods using high-performance liquid chromatography. Analyst. 105 (1980) 589–99. Retrieved 13 December 2016, from http://www.ncbi.nlm.nih.gov/pubmed/7416479

[26] F. Delgado-Vargas, O. Paredes-Lopez. Pigments as food colorants. In: CRC Press (Ed.), Natural colorants for food and nutraceutical uses. New York: CRC Press; 2003, pp. 35–62.

[27] Anvisa – Resolução – CNNPA n° 44, de 1977 (n.d.) Retrieved 13 December 2016, from http://www.anvisa.gov.br/anvisalegis/resol/44_77.htm

[28] S. Dixit, R.C. Pandey, M. Das, S.K. Khanna. Food quality surveillance on colours in eatables sold in rural markets of Uttar Pradesh. J. Food Sci. Technol. 32 (1995) 373–376.

[29] B.P. Harp, J.N. Barrows. US regulation of color additives in foods. In: M.J. Scotter editors. Colour additives for foods and beverages. Oxford: Woodhead Publishing; 2015, p. 75–88. doi:10.1016/B978-1-78242-011-8.00004-0.

[30] E. Stich. Food color and coloring food: quality, differentiation and regulatory requirements in the European Union and the United States. In: R.C. R. Schweiggert editors. Handbook on natural pigments in food and beverages.1st. Oxford: Woodhead Publishing; 2016, p. 3–27. doi:10.1016/B978-0-08-100371-8.00001-4.

[31] J.B. Hallagan, D.C. Allen, J.F. Borzelleca. The safety and regulatory status of food, drug and cosmetics colour additives exempt from certification. Food Chem. Toxicol. 33 (1995) 515–528. doi:10.1016/0278-6915(95)00010-Y.

[32] M.A. Prado, H.T. Godoy. Teores de corantes artificiais em alimentos determinados por cromatografia líquida de alta eficiência. Quim. Nova. 30 (2007) 268–273. doi:10.1590/S0100-40422007000200005.

[33] A.G. Fogg, K.S. Yoo. Direct differential-pulse polarographic determination of mixtures of the food colouring matters tartrazine-Sunset Yellow FCF, tartrazine-Green S and amaranth-Green S in soft drinks. Analyst. 104 (1979) 723–729. doi:10.1039/an9790400723.

[34] A.G. Fogg, K.S. Yoo. Direct differential-pulse polarographic determination of mixtures of food colouring matters, chocolate brown HT, tartrazine and Green S. Analyst. 104 (1979) 1087–1090. doi:10.1039/an9790401087.

[35] A.G. Fogg, A.A. Barros, J.O. Cabral. Differential-pulse adsorptive stripping voltammetry of food and cosmetic synthetic colouring matters and their determination and partial

identification in tablet coatings and cosmetics. Analyst. 111 (1986) 831–835. doi:10.1039/an9861100831.

[36] F.B. Dominguez, F.G. Diego, J.H. Mendez. Determination of sunset yellow and tartrazine by differential pulse polarography. Talanta. 37 (1990) 655–658. doi:10.1016/0039-9140(90)80213-Y.

[37] P.L.L. López-de-Alba, L. López-Martínez, L.M.M. De-León-Rodríguez. Simultaneous determination of synthetic dyes tartrazine, Allura Red and Sunset Yellow by differential pulse polarography and partial least squares. A multivariate calibration method. Electroanalysis. 14 (2002). doi:10.1002/1521-4109(200202)14:3<197::AID-ELAN197>3.0.CO;2-N.

[38] S. Combeau, M. Chatelut, O. Vittori. Identification and simultaneous determination of Azorubin, Allura Red and Ponceau 4R by differential pulse polarography: application to soft drinks. Talanta. 56 (2002) 115–122. doi:10.1016/S0039-9140(01)00540-9.

[39] A.G. Fogg, D. Bhanot. Voltammetric determination of synthetic food colouring matters at a stationary glassy carbon electrode. Analyst. 105 (1980) 868–872. doi:10.1039/an9800500868.

[40] E. Desimoni, B. Brunetti, M.S. Cosio. Determination of Patent Blue V (E131) at a nafion-modified glassy carbon electrode. Electroanalysis. 18 (2006) 231–235. doi:10.1002/elan.200503388.

[41] D.S. Nayak, N.P. Shetti. A novel sensor for a food dye erythrosine at glucose modified electrode. Sens. Actuators B: Chem. 230 (2016) 140–148. doi:10.1016/j.snb.2016.02.052.

[42] Y. Zhang, X. Zhang, X. Lu, J. Yang, K. Wu. Multi-wall carbon nanotube film-based electrochemical sensor for rapid detection of Ponceau 4R and Allura Red. Food Chem. 122 (2010) 909–913. doi:10.1016/j.foodchem.2010.03.035.

[43] P. Sierra-Rosales, C. Toledo-Neira, J.A. Squella. Electrochemical determination of food colorants in soft drinks using MWCNT-modified GCEs. Sens. Actuators B: Chem. 240 (2017) 1257–1264. doi:10.1016/j.snb.2016.08.135.

[44] Hair Color with L'Oreal (n.d.) Retrieved 19 December 2016, from http://www.hairproducts.com/articles/hair-color-with-loreal.html

[45] J.P. Thyssen, J.M.L. White. Epidemiological data on consumer allergy to p-phenylenediamine. Contact Dermatitis. 59 (2008) 327–343.

[46] Inmetro-Tinturas para Cabelo (n.d.) Retrieved 19 December 2016, from http://www.inmetro.gov.br/consumidor/produtos/tintura_cabelo.asp

[47] R.A.G. Oliveira, T.B. Zanoni, G.G. Bessegato, D.P. Oliveira, G.A. Umbuzeiro, M.V.B. Zanoni. A quìmica e toxicidade dos corantes de cabelo. Quim. Nov. xy (2014) 1–10.

[48] K.N. Jallad, C. Espada-Jallad. Lead exposure from the use of *Lawsonia inermis* (Henna) in temporary paint-on-tattooing and hair dying. Sci. Total Environ. 397 (2008) 244–250. doi:10.1016/j.scitotenv.2008.02.055.

[49] G.J. Nohynek, R. Fautz, F. Benech-Kieffer, H. Toutain. Toxicity and human health risk of hair dyes. Food Chem. Toxicol. 42 (2004) 517–543.

[50] C. Bolduc, J. Shapiro. Hair care products: Waving, straightening, conditioning, and coloring. Clin. Dermatol. 19 (2001) 431–436.

[51] K.C. Brown. Hair colouring. In: D.H. Johnson editors. Hair and Hair Care. 1st. New York: Marcel Dekker; 1977, p. 191–215.

[52] Y. Masukawa. Separation and determination of basic dyes formulated in hair care products by capillary electrophoresis. J. Chromatogr. A. 1108 (2006) 140–144. doi:10.1016/j.chroma.2006.01.007.

[53] J. Araldi, S.S. Guterres. Tinturas capilares: existe rico de câncer relacionado à utilização desses produtos? Infarma. 17 (2005) 78–83.

[54] A. Chisvert, A. Cháfer, A. Salvador. Hair dyes in cosmetics. Regulatory aspects and analytical methods. In: A.Salvador, A. Chisvert editors. Analysis of cosmetic products.1st. Valencia: Elsevier; 2007, p. 190–209. ISBN 9780080475318.

[55] A. Guerra-Tapia, E. Gonzalez-Guerra. Hair cosmetics: dyes. Actas Dermosifiliogr. 105 (2014) 833–839.

[56] G.J. Nohynek, E. Antignac, T. Re, H. Toutaina. Safety assessment of personal care products/cosmetics and their ingredients. Toxicol. Appl. Pharmacol. 243 (2010) 239–259.

[57] T. Menné, H. Sosted, S.C. Rastogi, K.E. Andersen, J.D. Johansen. Hair dye contact allergy: quantitative exposure assessment of selected products and clinical cases. Contact Dermatitis. 50 (2004) 344–348.

[58] K. Yazar, A. Boman, C. Lidén. p-Phenylenediamnine and other hair dye sensitizers in Spain. Contact Dermatitis. 66 (2011) 27–32.

[59] J.P. Thyssen, K.E. Andersen, M. Bruze, T. Diepgen, A.M. Giménez-Arnau, M. Gonçalo, A. Goossens, C.L. Coz, J. Mcfadden, T. Rustemeyer, I.R. White, J.M. White, J.D. Johansen. p-Phenylenediamine sensitization is more prevalent in central and southern european patch test centres than in Scandinavian: results from a multicenter study. Contact Dermatitis. 60 (2009) 314–319.

[60] B.N. Ames, H.O. Kammen, E. Yamasaki. Hair dyes are mutagenic: identification of a variety of mutagenic ingredients. Proc. Natl. Acad. Sci. U. S. A. 72 (1975) 2423–2427. Retrieved 20 December 2016, from http://www.ncbi.nlm.nih.gov/pubmed/1094469

[61] T.B. Zanoni, F. Hudari, A. Munnia, M. Peluso, R.W. Godschalk, M.V.B. Zanoni, G.J.M. den Hartog, A. Bast, S.B.M. Barros, S.S. Maria-Engler, G.J. Hageman, D.P. de Oliveira. The oxidation of p-phenylenediamine, an ingredient used for permanent hair dyeing purposes, leads to the formation of hydroxyl radicals: oxidative stress and DNA damage in human immortalized keratinocytes. Toxicol. Lett. 239 (2015) 194–204. doi:10.1016/j.toxlet.2015.09.026.

[62] J.M.L. White, P. Kullavanijaya, I. Duangdeeden, R. Zazzeroni, N.J. Gilmour, D.A. Basketter, J.P. McFadden. p-Phenylenediamine allergy: the role of Bandrowski's base. Clin. Exp. Allergy. 36 (2006) 1289–1293. doi:10.1111/j.1365-2222.2006.02561.x.

[63] J. Farrell, C. Jenkinson, S.N. Lavergne, J.L. Maggs, B. Kevin Park, D.J. Naisbitt. Investigation of the immunogenicity of p-phenylenediamine and Bandrowski's base in the mouse. Toxicol. Lett. 185 (2009) 153–159. doi:10.1016/j.toxlet.2008.12.008.

[64] C. Olson, H.Y. Lee, R.N. Adams. Polarographic oxidation and reduction of p-nitroaniline at carbon paste electrodes. J. Electroanal. Chem. 2 (1961) 396–399. doi:10.1016/0022-0728(61)85020-1.

[65] K. Bratin, P.T. Kissinger, R.C. Briner, C.S. Bruntlett. Determination of nitro aromatic, nitramine, and nitrate ester explosive compounds in explosive mixtures and gunshot residue by liquid chromatography and reductive electrochemical detection. Anal. Chim. Acta. 130 (1981) 295–311. doi:10.1016/S0003-2670(01)93007-7.

[66] L.K.J. Tong, K. Liang, W.R. Ruby. Application of rotating disk electrodes to the study of electrode-initiated reactions of p-phenylenediamines. J. Electroanal. Chem. Interfacial Electrochem. 13 (1967) 245–262. doi:10.1016/0022-0728(67)80122-0.

[67] N.S. Lawrence, E.L. Beckett, J. Davis, R.G. Compton. Voltammetric investigation of hair dye constituents: application to the quantification of p-phenylenediamine. Analyst. 126 (2001) 1897–1900. doi:10.1039/b104641c.

[68] R.A.G. Oliveira, M.V.B. Zanoni. Highly ordered TiO_2 nanotubes for electrochemical sensing of hair dye basic brown 17. Electroanalysis. 25 (2013) 2507–2514. doi:10.1002/elan.201300322.

[69] P. Zhao, J. Hao. 2,6-diaminopyridine-imprinted polymer and its potency to hair-dye assay using graphene/ionic liquid electrochemical sensor. Biosens. Bioelectron. 64 (2015) 277–284.

[70] G.T. Corrêa, A.A. Tanaka, M.I. Pividori, M.V.B. Zanoni. Use of a composite electrode modified with magnetic particles for electroanalysis of azo dye removed from dyed hair strands. J. Electroanal. Chem. 782 (2016) 26–31. doi:10.1016/j.jelechem.2016.09.044.

[71] J.P. Hart, S.A. Wring. Recent developments in the desing and application of screen-printed electrochemical sensor for biomedical, environmental and industrial analysis. Trends Anal. Chem. 16 (1997) 89–103.

[72] V.B. Nascimento, L. Angnes. Eletrodo fabricados por "silk-screen". Quim. Nova. 21 (2008) 614–629.

[73] S. Eissa, C. Tlili, L. L'hocine, M. Zourob. Electrochemical immunosensor for the milk allergen β-lactoglobulin based on electrografting of organic film on graphene modified screen-printed carbon electrodes. Biosens. Bioelectron. 38 (2012) 308–313. doi:10.1016/j.bios.2012.06.008.

[74] A. Bafana, S.S. Devi, T. Chakrabarti. Azo dyes: past, present and the future. Environ. Rev. 19 (2011) 350–371. doi:10.1139/a11-018.

[75] P. Seesuriyachan, S. Takenaka, A. Kuntiya, S. Klayraung, S. Murakami, K. Aoki. Metabolism of azo dyes by *Lactobacillus casei* TISTR 1500 and effects of various factors on decolorization. Water Res. 41 (2007) 985–992. doi:10.1016/j.watres.2006. 12 .001.

[76] C.C.I. Guaratini, M.V.B. Zanoni. Corantes têxteis. Quim. Nova. 23 (2000) 71–78.

[77] K. Hunger. Industrial dyes: chemistry, properties, applications. Weinheim: Wiley-VCH Verlag GmbH & Co. KGaA, FRG; 2004. doi:10.1002/3527602011.

[78] E. Forgacs, T. Cserháti, G. Oros. Removal of synthetic dyes from wastewaters: a review. Environ. Int. 30 (2004) 953–971. doi:10.1016/j.envint.2004.02.001.

[79] A.A. Vaidya, K.V. Datye. Environmental pollution during chemical processing of synthetic fibers. Colourage. 14 (1982) 3–10. Retrieved 11 January 2017, from http://www.sciepub.com/reference/35260

[80] C.C.I. Guaratini, A.G. Fogg, M.V.B. Zanoni. Studies of the voltammetric behavior and determination of diazo reactive dyes at mercury electrode. Electroanalysis. 13 (2001) 1535–1543. doi:10.1002/1521-4109(200112)13:18<1535::AID-ELAN1535>3.0.CO;2-H.

[81] A. Radi, M.R. Mostafa, T.A. Hegazy, R.M. Elshafey. Electrochemical study of vinylsulphone azo dye Reactive Black 5 and its determination at a glassy carbon electrode. J. Anal. Chem. 67 (2012) 890–894.

[82] A.-E. Radi, H.M. Nassef, A. El-Basiony. Electrochemical behavior and analytical determination of Reactive Red 231 on glassy carbon electrode. Dye. Pigment. 99 (2013) 924–929. doi:10.1016/j.dyepig.2013.07.025.

[83] M.E. Osugi, P.A. Carneiro, M.V.B. Zanoni. Determination of the phthalocyanine textile dye, reactive turquoise blue, by electrochemical techniques. J. Braz. Chem. Soc. 14 (2003) 660–665. doi:10.1590/S0103-50532003000400025.

[84] Y.M. Park, Y.H. Kim, T. Yamamoto. Determination of dye concentration in water using mesoporous particle coated QCR sensor. Sens. Actuators B: Chem. 125 (2007) 468–473. doi:10.1016/j.snb.2007.02.042.

[85] D.P. Santos, M.A.G. Trindade, R.A.G. Oliveira, M.E. Osugi, A.R. Bianchi, M.V.B. Zanoni. Electrochemical method for quantitative determination of trace amounts of disperse dye in wastewater. Color. Technol. 130 (2014) 43–47. doi:10.1111/cote. 12059.

[86] Z. Liu, H. Zhai, Z. Chen, Q. Zhou, Z. Liang, Z. Su. Simultaneous determination of Orange G and Orange II in industrial wastewater by a novel Fe_2O_3/MWCNTs-COOH/OP modified carbon paste electrode. Electrochim. Acta. 136 (2014) 370–376.

[87] P. Manisankar, G. Selvanathan, S. Viswanathan, H. Gurumallesh Prabu. Electrochemical determination of some organic pollutants using wall-jet electrode. Electroanalysis. 14 (2002) 1722–1727. doi:10.1002/elan.200290016.

[88] A. Chen, E.I. Rogers, R.G. Compton. Abrasive stripping voltammetry in room temperature ionic liquids. Electroanalysis. 21 (2009) 29–35. doi:10.1002/elan.200804401.

[89] A. Vuorema, P. John, M. Keskitalo, F. Marken, A. Philip, J. Ae, M. Keskitalo, F. Marken. Electrochemical determination of plant-derived leuco-indigo after chemical reduction by glucose. J. Appl. Electrochem. 38 (2008) 1683–1690. doi:10.1007/s10800-008-9617-0.

[90] S.S.M. Hassan, A.H. Kamel, H.A. El-Naby. Flow-through potentiometric sensors for alizarin red S dye and their application for aluminum determination. J. Chin. Chem. Soc. 61 (2014) 295–302. doi:10.1002/jccs.201300293.

[91] R.B. Orelup. Colored petroleum markers. US Patent 4735631 A, 1988.

[92] M.R. Friswell, M.P. Hinton. Markers for petroleum, method of tagging, and method of detection. US Patent 5205840 A, 1993.

[93] S. Suwanprasop, T. Nhujak, S. Roengsumran, A. Petsom. Petroleum marker dyes synthesized from cardanol and aniline derivatives. Ind. Eng. Chem. Res. 43 (2004) 4973–4978. doi:10.1021/IE030739S.

[94] A.V. Nowak. 4918020, 1990. Retrieved 11 January 2017, from https://www.google.com.br/_/chrome/newtab?espv=2&ie=UTF-8

[95] European Refining and Marketing. Fuels Refining and Marketing in Europe and The Former Soviet Union. 1 (2002) 1–32.

[96] M. A.G. Trindade, N.R. Stradiotto, M. V.B. Zanoni. Corantes marcadores de combustíveis: legislação e métodos analíticos para detecção. Quim. Nov. 34 (2011) 1683–1691.

[97] T. Linsinger, G. Koomen, H. Emteborg, G. Roebben, G. Kramer, A. Lamberty. Validation of the European Union's reference method for the determination of Solvent Yellow 124 in gas oil and kerosene. Energy Fuels. 18 (2004) 1851–1854. doi:10.1021/EF049820D.

[98] E.C. Teixeira, S. Feltes, E.R.R. de Santana. Estudo das emissões de fontes móveis na região metropolitana de Porto Alegre, Rio Grande do Sul. Quim. Nova. 31 (2008) 244–248. doi:10.1590/S0100-40422008000200010.

[99] G.N.E.B. AGÊNCIA NACIONAL DO PETRÓLEO, RESOLUÇÃO ANP N° 36, DE 6.12.2005-DOU 7.12.2005 (n.d.) Retrieved 11 January 2017, from http://www.puntofocal.gov.ar/notific_otros_miembros/bra199_t.pdf

[100] M.A.G. Trindade, M.V. Boldrin Zanoni. Square-wave voltammetry applied to the analysis of the dye marker, solvent blue 14, in kerosene and fuel alcohol. Electroanalysis. 19 (2007) 1901–1907. doi:10.1002/elan.200703964.

[101] M.A.G. Trindade, M.V.B. Zanoni. Voltammetric sensing of the fuel dye marker solvent blue 14 by screen-printed electrodes. Sens. Actuators B: Chem. 138 (2009) 257–263. doi:10.1016/j.snb.2009.01.043.

[102] M.A.G. Trindade, V.S. Ferreira, M.V.B. Zanoni. A square-wave voltammetric method for analysing the colour marker quinizarine in petrol and diesel fuels. Dye. Pigment. 74 (2007) 566–571. doi:10.1016/j.dyepig.2006.03.020.

[103] M.A.G. Trindade, U. Bilibio, M.V.B. Zanoni. Enhancement of voltammetric determination of quinizarine based on the adsorption at surfactant-adsorbed-layer in disposable electrodes. Fuel. 136 (2014) 201–207. doi:10.1016/j.fuel.2014.07.044.

Chapter 5

Applications of Voltammetric Analysis to Wine Products

Dolores Hernanz-Vila, M. José Jara-Palacios,
M. Luisa Escudero-Gilete and Francisco J. Heredia

Additional information is available at the end of the chapter

http://dx.doi.org/10.5772/67696

Abstract

Wine contains polyphenols that are responsible for its quality. Moreover, phenolic compounds have antioxidant properties and benefits on human health. Cyclic voltammetry (CV) was the first electrochemical method used for polyphenols characterization and determination of polyphenols content in wine products. Electrochemical behaviour of standard solutions of phenolic compounds has been investigated and evaluated the importance of the phenolic concentration and pH. The electrochemical parameters extracted from the voltammograms have been correlated with the antioxidant potential in wine products. In addition, CV allowed establishing differences in the antioxidant activity of wines with different addition of grape seeds. In winemaking by-products, different I_{pa} and Q_{500} values were found depending on the state of maturation of the grape pomace. On the other hand, the total flavonoids and phenolic acids contents were significantly correlated to the electrochemical parameters. Differences for the electrochemical parameters were found between by-products, being pomace and seeds which presented the greatest values of Q_{500}. Simple regression analyses showed that voltammetric parameters are correlated to their values of lipid peroxidation inhibition by thiobarbituric acid reactive substances method. Our results open the possibility of CV as a promising technique to estimate the global antioxidant potential of wine products rich in phenolic compounds.

Keywords: antioxidant activity, phenolic composition, electrochemical parameters, wine, wine by-products

1. Introduction

Grape juice is rich in phenolic compounds and plays a very important role in winemaking, mainly due to its content in pigments and tannins. Several hundred phenolic compounds have been identified in grapes, and they are transferred from grapes to wines during vinification. These compounds accumulate rapidly during berry maturation and they are very important for grape (and wine) character because they include red pigments, astringent flavours and browning substrates [1, 2].

The total phenol content of wine is less than that present in the grape because traditional methods of destemming, crushing and fermenting usually give extraction rates of no more than 60% [3].

The major phenolic compounds found in wine are either members of the diphenylpropanoids (flavonoids) or phenylpropanoids (nonflavonoids). Flavonoids, including anthocyanins, flavonols and flavan-3-ols (catechin, epicatechin, and their procyanidin polymers), are the most important phenolics for wine quality [4]. Flavonoids are derived primarily from the seeds, skins and stems of the grape. Anthocyanins and flavonols are extracted mainly from skins, and catechins and leucoanthocyanins reside mainly in seeds and stems. Increasing skin contact time and fermentation temperature, and the degree of berry disruption increase the flavonoid content of a wine [5].

Nonflavonoids are structurally simpler, but their origin in wine is more diverse. In wines not aged in oak, the primary nonflavonoids are derivatives of hydroxycinnamic and hydroxybenzoic acids [1]. They are stored primarily in cell vacuoles of skin and pulp, and are easily extracted on crushing. The most numerous and variable are hydroxycinnamic acid derivatives. They occur principally as esters with tartaric acid (for example, caftaric, coutaric and fertaric acids, the tartaric acid esters of caffeic, p-coumaric and ferulic acids, respectively), but may also be associated with sugars, various alcohols or other organic acids. The esters also slowly hydrolyse during fermentation [6]. The most common nonflavonoid in grapes is caftaric acid, one of the primary substrates for polyphenol oxidase. It often plays an important role in oxidative browning of must [7, 8].

Particularly in wine red, flavonoids and some phenolic acids (colourless phenols) are involved in the chemical stabilization of anthocyanin pigments by means of non-covalent interactions through intermolecular co-pigmentation reactions [8, 9]. Studies were carried out in model solution and focused on the application of objective colour measurements. These studies have demonstrated that co-pigmentation causes the stabilization of the coloured forms of the anthocyanins and consequently enhance their colour of wine [10, 11].

In the last years, the pre-fermentative cold maceration, also known as cold soaking or cryomaceration, is being increasingly used by enologists worldwide in order to improve some important quality characteristics of wines such as colour and aroma [12–16]. This technique consists in maintaining the crushed grapes at low temperatures (5–10°C) for a variable period (from one to several weeks), and thus the beginning of the fermentation process is delayed. During this period, the extraction of polyphenols from the skins to the must takes place in the absence of ethanol. With reference to the phenolic compounds, these compounds contribute to colour stability since they can act as oxidation substrates in white wines [17–19]. Controlling skin contact conditions is vital to obtain high-quality white wines [20]. Skin contact may greatly increase both the total hydroxycinnamate and flavanol concentration [21].

Wine, especially red wine, is a very rich source of polyphenols, such as flavanols (catechin, epicatechin, etc.), flavonols (quercetin, rutin, myricetin, etc.), anthocyanins (the most abundant is malvidin-3-O-glucoside), oligomeric and polymeric proanthocyanidins, phenolic acids (gallic acid, caffeic acid, p-coumaric acid, etc.), stilbenes (trans-resveratrol) and many others polyphenols.

Winemaking generates a high amount of by-products that cause environmental and economic problems, which could be minimized by the exploitation and valorization of those products, such as their use in pharmaceutical and food industries. Grape pomace, consisting of seeds, skins and stems, is the main winemaking by-product and is a rich source in phenolic compounds with interest by their biological and antioxidant properties [22, 23].

Many of these phenolic compounds have been reported to have multiple biological activities, including cardioprotective, anti-inflammatory, anti-carcinogenic, antiviral and antibacterial properties [24, 25]. These biological properties are attributed mainly to their powerful antioxidant and antiradical activity.

1.1. Antioxidant activity

Oxidative stress takes place when there is an imbalance between the production of reactive species and the antioxidant defence systems, that is, there is a disturbance in the pro-oxidant-antioxidant balance in favour of the oxidant species, leading to potential damage [26–29]. The oxidative damage that takes place is defined as biomolecular damage caused by the attack of reactive species on the cells and tissues of the living organisms [30].

These reactive species are oxidant agents and/or convertible into free radicals easily. Reactive oxygen species (ROS) and reactive nitrogen species (RNS) are the most common reactive species. ROS, such as superoxide anion radical ($O_2^{\bullet -}$), singlet oxygen ($^1O_2$), hydrogen peroxide (H_2O_2) and hydroxyl radical ($OH^{\bullet}$), are constantly generated in living organisms by endogenous (metabolism, inflammatory reactions) or exogenous sources (environmental factors) [30].

The action mechanism of the free radical is based on the attack to the target molecule to remove a hydrogen atom, or an electron, and so the unpaired electron of the radical one turns into a more stable electrons par. In this process, the target molecule oxidizes. The principal target molecules are DNA, proteins and lipids. Antioxidants are compounds or systems that can safely interact with free radicals and terminate the chain reaction before vital molecules are damaged.

The antioxidant capacity is defined as the ability of compound (or mixture of compounds) to inhibit the oxidative degradation of various compounds. Antioxidant functions imply lowering oxidative stress, DNA mutations, malignant transformations, as well as other parameters of cell damage.

The human organism has an antioxidant defence system to neutralize the excessive levels of ROS and RNS and protect to the cells from oxidative damage. This defence system can be from endogenous (enzymatic and non-enzymatic) or exogenous origin. The enzymatic systems, especially superoxide dismutases (SOD), catalases (CAT) and glutathione peroxidases (GPX), are recognized as being highly efficient in ROS detoxification [31–33]. The main small molecule non-enzymatic antioxidants present in the human organism are bilirubin, estrogenic sex hormones, uric acid, ascorbic acid, coenzyme Q, melanin, melatonin, α-tocopherol and lipoic acid [34, 35]. The diet provides to the body with basic nutrients (proteins, vitamins, minerals) and phytochemical substances that have antioxidant activity and help endogenous defence systems.

Fruits and vegetables are accepted as good sources of natural antioxidants, which provide protection against free radicals and have been associated with lower incidence and mortality rates of cancer and heart diseases in addition to a number of other health benefits [36–38]. Higher plants and their constituents provide a rich source of natural antioxidants, such as carotenoids, tocopherols and polyphenols that are found abundantly in spices, herbs, fruits, vegetables, cereals, grains, seeds, teas and oils. In addition, by-products from the food and agricultural industries have been explored for their potential use as antioxidants. For example, seeds, skins and stem of pomace from winemaking, hulls, shells and skins of nuts and cereals, citrus peels and seeds have been found to possess antioxidant activity [39–42].

There are several methods for evaluating antioxidant activity, either *in vitro* or *in vivo*. The *in vitro* assays can be classified into chemical methods (spectrophotometric methods and electrochemical techniques) and biological methods (cellular systems).

The most used spectrophotometric methods are 3-ethylbenzothiazoline-6-sulfonic acid (ABTS), 1,1-diphenyl-2-picrylhydrazyl (DPPH) and oxygen radical absorbance capacity (ORAC), which measure the ability of antioxidants to scavenge a radical; ferric reducing antioxidant power (FRAP), cupric reducing antioxidant capacity (CUPRAC) and cerium reducing antioxidant capacity (CERAC), which measure the capacity of antioxidant to reduce metals; and thiobarbituric acid reactive substances (TBARS), which is used to measure the lipid peroxidation inhibition [43]. These spectrometric methods are mainly used in the analysis of antioxidant properties. However, these methods are dependent on many parameters, such as temperature, time of the analysis, character of a compound or mixture of compounds (extracts), concentration of antioxidants and pro-oxidants and many other substances. In addition, they are based on different action mechanisms so the antioxidant activity value of the samples differs according to the used test [44, 45]. The measurement of antioxidant activity cannot be evaluated satisfactorily using a simple antioxidant test due to the many variables influencing the results [46].

Total phenolic content (TPC) is another parameter used for evaluation of antioxidant extracts. The Folin-Ciocalteu assay is the well-known method for determination of TPC. This method is used to analyses phenolic components in wine, and it became a routine analysis for antioxidant assessment of food and plant extracts [43].

On the other hand, electrochemical methods for evaluating antioxidant activity have emerged in the past decade, among which cyclic voltammetry (CV) has attracted much attention as an alternative method to conventional chemical assays. Cyclic voltammetry measures electron-donation capability (redox potential) of antioxidants, which respond to a voltammetric scan according to their redox potential [43].

1.2. Cyclic voltammetry

CV is a simple, fast and inexpensive electrochemical technique that could become an alternative to traditional spectrophotometric techniques to measure the antioxidant activity. CV has already been applied to evaluate of antioxidant capacity in blood plasma [47], plant extracts [48], vegetable oils [49], milk [50] and orange juice [51].

CV has been successfully used to determine the phenolic content of wines and to correlate the analytical response to the antioxidant capacity of these wines [52–54]. It was shown that CV provides a qualitative and quantitative assessment of wine phenolics based on their reducing strength, and charge passed to 500 mV (vs Ag/AgCl). Other studies also demonstrated the coherence of the cyclic voltammetric response with the information provided by HPLC, Folin-Ciocalteu assays and absorbance at 280 nm on white and red wines [53]. CV does not allow identifying individual antioxidants present in the sample; however, the technique provides the sum of total antioxidants [55]. According to Kilmartin et al. [52], Makhotkina and Kilmartin [56] and Rebelo et al. [57], each voltammetric peak is ascribed to different groups of phenolic compounds.

In addition, CV was also used to investigate the influence of sulphur dioxide, glutathione and ascorbic acid on polyphenol oxidation processes relevant to wine oxidation [58] and to correlate analytical response to sensory characteristics such as astringency [59].

2. Experimental part

2.1. Cyclic voltammetry measurement

The experimental configuration for recording cyclic voltammograms consists of an electrochemical cell that has a tree electrodes, counter or auxiliary electrode, reference electrode and working electrode, all immersed in a liquid and connected to a potentiostat. The potentiostat AUTOLAB model PGSTAT 302 N (Metrohm-Eco Chemie, Netherlands), controlled by a General Purpose Electrochemical System (GPES) software and conventional three-electrode system consisting of a glassy carbon working electrode, platinum auxiliary electrode and Ag/AgCl reference electrode, was used for all electrochemical measurements.

All measurements were carried out at room temperature using a conventional three-electrode system. Prior to the measurements, the working electrode was polished in alumina/water suspension, rinsed with Milli-Q water and sonicated for 2 min. The electrolyte solution was transferred into a glass water-jacketed electrochemical cell (EG&G, Princeton, NJ) connected to a circulator that held the sample temperature at 25.0 ± 0.5°C. The solution was de-aerated with an inert gas (N_2) for 10 min, and after a 1 min running scan was taken. The cyclic voltammogram scans were made from 0.0 to 0.5 V at a scanning rate of 5 mV/s for winemaking by-products and 0.0 to 1.0 V for wine solutions.

The electrochemical parameters extracted from the cyclic voltammetry curve were the peak anodic current and potential ($I_{p,a}$ and $E_{p,a}$, respectively), the peak cathodic current and potential ($I_{p,c}$ and $E_{p,c}$, respectively), the potential mid-way between the anodic and cathodic peaks (E_{mid}) calculated from ½ ($E_{p,c} + E_{p,a}$), $I_{p,a}/I_{p,c}$ and $E_{p,a} - E_p/2$. The anodic current area (Q), which represents the total integrated area of the cyclic voltammogram for scans taken from 0 to 1 V (QT) or from 0 to 0.5 V (Q_{500}), was also extracted. In addition, QI, QII and QIII were also calculated; these parameters represent the area corresponding to peaks I, II and III, respectively (**Figure 1**).

These parameters were taken from the cyclic voltammograms after subtracting cyclic voltammogram data of the blank (1 mL of 75% methanol diluted with 25 mL phosphate buffer). All of the cyclic voltammograms were recorded in triplicate.

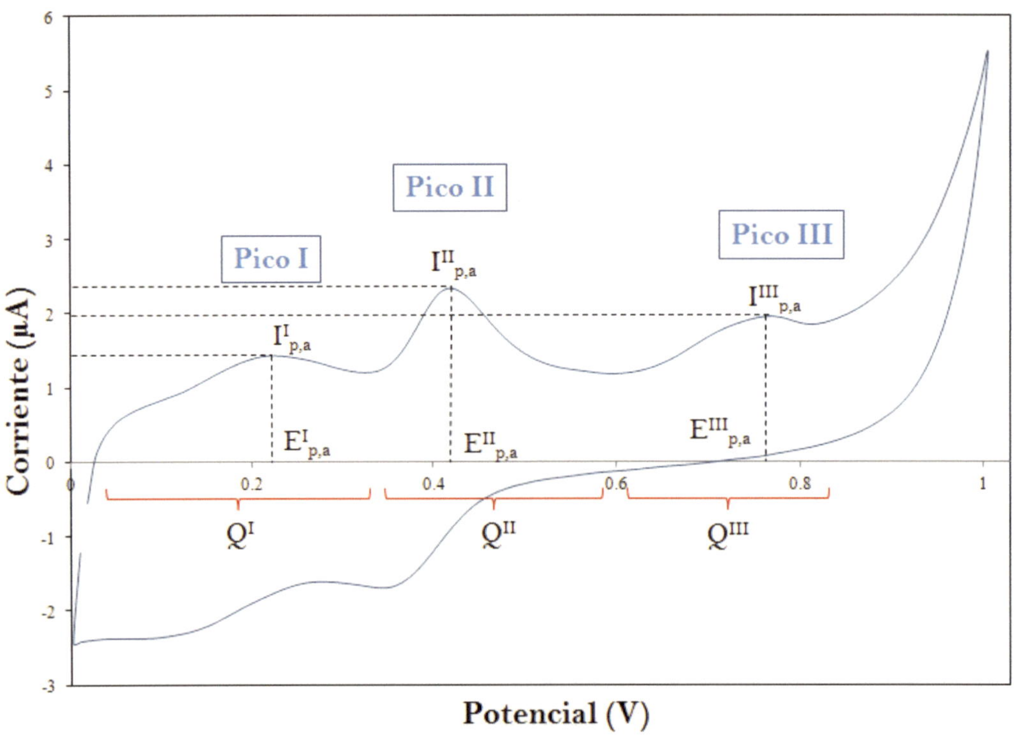

Figure 1. Representative cyclic voltammogram of a grape pomace.

http://dx.doi.org/10.5772/67696

2.2. Applications of cyclic voltammetry

2.2.1. *Electrochemical properties of phenolic compounds*

In this study, CV was used to monitor the electrochemical behaviour of four phenolic compounds representing the main phenolic groups found in wines and winemaking by-products, namely gallic acid (hydroxibenzoic acid), caffeic acid (hydroxycinnamic acid), catechin (flavanol) and quercetin (flavonol). These compounds have in common a catechol moiety believed to be the electrochemically active group [52]. All the compounds measured acted as powerful antioxidants and were oxidized on a glassy carbon electrode. The electrochemical measurements were taken at different pH (3.6 and 7), concentrations (100 and 500 mg/L) and potential scans (0.5 and 1 V). **Table 1** shows the electrochemical parameters ($I_{p,a}$, $E_{p,a}$, Q) extracted from the cyclic voltammetry curves of the compounds. In a previous paper, Q and $I_{p,a}$ parameters showed a strong and significant correlations ($R = 0.95$, $p < 0.05$), and therefore results are discussed based on parameter $I_{p,a}$.

The pH influence was very important on electrochemical behaviour of these compounds. As can be seen in **Figure 2**, cyclic voltammograms of phenolic compounds, adjusted to pH 7, showed one well-defined anodic peak between 0.2 and 0.3 V. However, two anodic peaks were exhibited to pH 3.6, between 0.1 and 0.3 V, and between 0.3 and 0.5 V. Considering cyclic voltammograms to pH 7, catechin and quercetin showed higher values of $I_{p,a}$ (1.32 μA) than gallic and caffeic acids (1.20 and 1.10 μA, respectively). The anodic peak potential ($E_{p,a}$) ranged between 0.25 V for quercetin and 0.27 for caffeic acid.

			Peak I			Peak II		
	pH	**Concentration (mg/L)**	E_{pa} **(V)**	I_{pa} **(μA)**	Q	E_{pa} **(V)**	I_{pa} **(μA)**	Q
Gallic acid	3	100	0.21	1.76	0.346	0.43	4.24	0.489
	3	500	0.22	1.93	0.354	0.43	5.97	0.659
	7	100	0.26	1.20	-	-	-	-
Caffeic acid	3	100	0.22	1.94	0.371	0.46	3.86	0.446
	3	500	0.21	1.74	0.330	0.46	5.25	0.569
	7	100	0.27	1.10	-	-	-	-
Catechin	3	100	0.21	1.72	0.341	0.46	3.29	0.415
	3	500	0.24	1.61	0.306	0.41	2.59	0.333
	7	100	0.26	1.32	-	-	-	-
Quercetin	3	100	0.22	1.88	0.363	0.45	2.52	0.340
	3	500	0.22	1.73	0.328	0.42	2.49	0.329
	7	100	0.25	1.32	-	-	-	-

Table 1. Electrochemical parameters of anodic peaks extracted from the cyclic voltammetry curves of the phenolic compounds.

Regarding to cyclic voltammograms to pH 3.6, values of $I_{p,a}$, $E_{p,a}$ and Q depended on the peak (I or II) and the concentration (100 or 500 mg/L). As can be observed in **Table 1**, peak II had higher values of $I_{p,a}$ than peak I; gallic acid showed the highest values at 100 and 500 mg/L (4.24 and 5.97 μA, respectively).

It is known that pH is the most significant factor determining the antioxidant activity of phenolic compounds. Yakovleva et al. [48] observed that with increasing pH values the anodic peak voltage decreased, which was caused by the decrease in the degree of antioxidant protonation and the resulting shift of the charge of the molecule to negative values.

The phenolic concentration is also an important factor for values of electrochemical parameters. As previously described, $I_{p,a}$ increases with increasing concentrations of phenolics although the relationship is not always linear [53]. $I_{p,a}$ increases with the concentration for gallic acid (4.24 and 5.97 μA for 100 and 500 mg/L, respectively) and caffeic acid (3.86 and 5.25 μA for 100 and 500 mg/L, respectively); however, this increase did not occur for catechin and quercetin (**Table 1**).

Cyclic voltammograms of gallic acid and catechin at 1 V are shown in **Figure 3**. The main peak for both compounds was at 0.43 V, and $I_{p,a}$ was 6.98 and 2.95 μA for gallic acid and catechin, respectively. The potential range is a significant factor since the flexibility of adjusting the electrode potential to higher values allows a progressively wider range of phenolic compounds to be monitored [60].

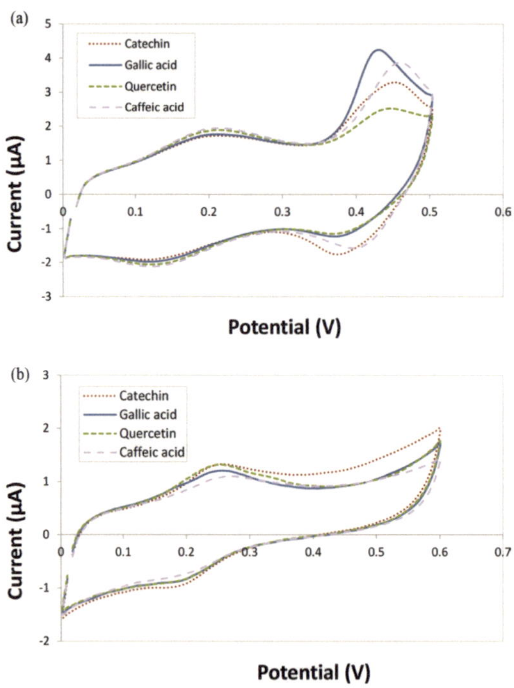

Figure 2. Cyclic voltammograms of phenolic compounds adjusted to pH 3.6 (a) and pH 7 (b).

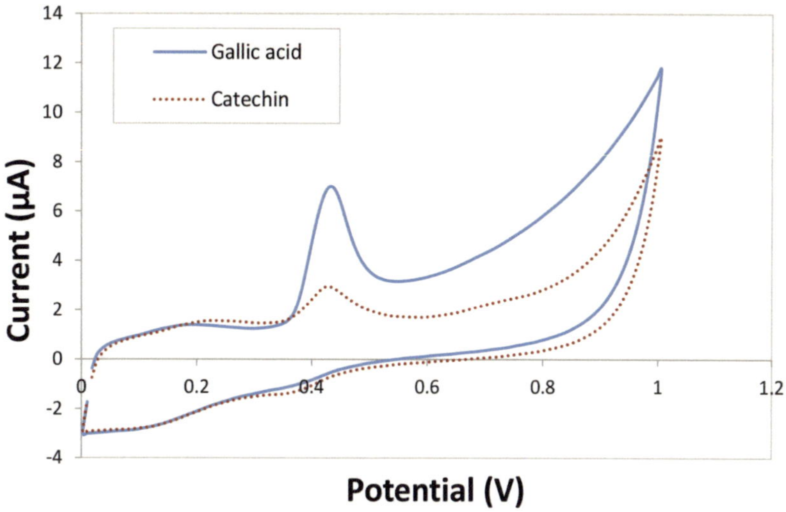

Figure 3. Cyclic voltammograms of gallic acid and catechin at 1 V.

2.2.2. Electrochemical properties of wine products

Before measures by CV, the samples containing phenolic compounds must be prepared correctly. It is important to control factors such as phenolic concentration and pH of the solution, because, as previously mentioned, these factors influence the electrochemical properties.

In order to obtain the suitable phenolic concentration and pH, samples are diluted with the corresponding buffer: 0.1 M sodium acetate-acetic acid buffer for pH 3.6, and 5% (w/v) 50 mM disodium hydrogen phosphate and 35% (w/v) 50 mM sodium dihydrogen phosphate for pH 7 [39, 61].

For this study, several dilutions (1/50, 1/25, 1/17, 1/12.5) of grape pomace extract were prepared in phosphate buffer (pH 7). **Table 2** shows the electrochemical parameters for diluted grape pomace extract determined from CV. These dilutions provided well-defined voltammetric peaks which had peak currents ($I_{p,a}$ and $I_{p,c}$) that changed linearly with wine dilution. $I_{p,a}$ ranged between 2.15 and 2.62 μA for most diluted (1/50) and most concentrated (1/12.5) solutions, respectively. As can be observed in **Figure 4**, the anodic current area was highest for solution most concentrated (Q_{500} = 5.67), which has more amount of phenolic compounds.

In order to explore the relationship between the concentration of grape pomace extract (according to dilution) and the electrochemical parameters ($I_{p,a}$ and Q_{500}), simple correlation analysis was realized. Significant and high linear correlations were found for $I_{p,a}$ and Q_{500} with concentration (**Figure 5**).

Each type of sample requires a different dilution to obtain well-defined voltammetric peaks and anodic peaks charge directly proportional to the volume fraction of the samples. Kilmartin [60] indicated that white wines require around 10-fold dilution and red wines up to 400-fold.

Rebelo et al. [57] reported that red wine samples required a 50-fold dilution to reach a range in which the anodic peak charge was directly proportional to the final volume of the wines. In our previous studies [39, 61, 62], a 25-fold dilution was required for winemaking by-products and wine samples.

Dilution	$E_{p,a}$ (mV)	$E_{p,c}$ (mV)	$I_{p,a}$ (μA)	$I_{p,c}$ (μA)	$E_{mid\ (mV)\ (E^{o\prime})}$	$E_{p,a}-E_{p/2}$	$I_{p,c}/I_{p,a}$	$\Delta E > 59$	Q_{500}
1/50	236	171	2.15	1.72	203	65	0.80	65	5.02
1/25	236	163	2.41	1.86	200	65	0.77	73	5.27
1/17	236	163	2.59	2.00	200	56	0.77	73	5.59
1/12.5	244	155	2.62	2.00	200	64	0.76	89	5.67

Table 2. Electrochemical parameters for diluted pomace extracts determined from cyclic voltammetry.

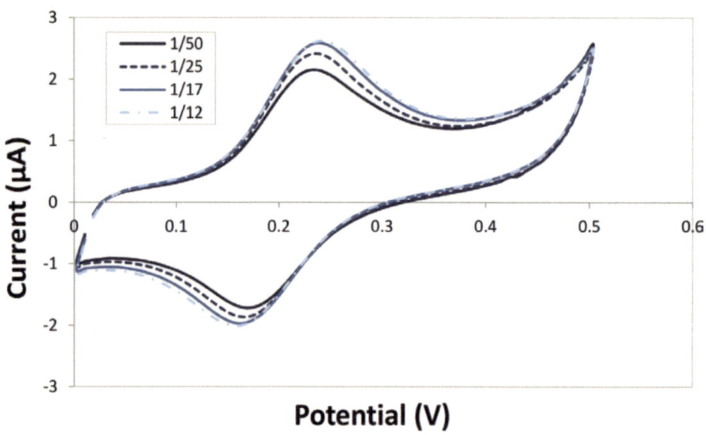

Figure 4. Cyclic voltammograms of diluted pomace extracts.

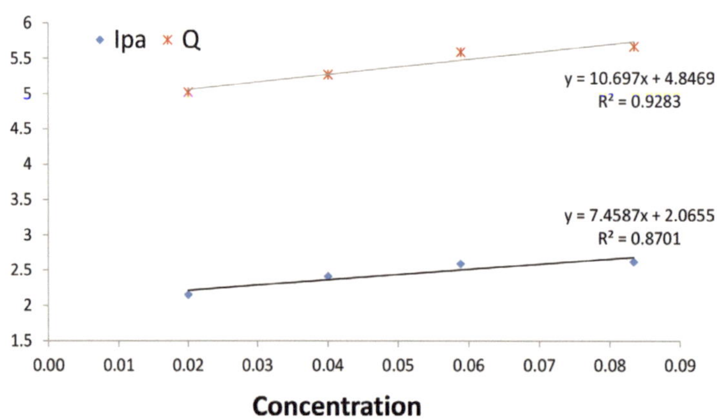

Figure 5. Correlations between the concentration of grape pomace extract and the electrochemical parameters ($I_{p,a}$ and Q_{500}).

2.2.2.1. *Winemaking by-products*

CV was used for determining the total antioxidant activity of phenolic compounds present in winemaking by-products.

Firstly, extracts from grape pomaces (including seeds, skins and stems) at different states of maturation (early, technological and late harvest: EH, TH, LH, respectively) were measured at pH 7 and 0.5 V, and a one well-defined anodic peak was showed at 0.24 V in the three cyclic voltammograms (**Figure 6**). Differences were found in I_{pa} depending on the state of maturation (**Table 3**). $I_{p,a}$ was highest for grape pomace at EH, followed by TH and LH (1.62, 1.50 and 1.28 μA, respectively). In addition, the anodic current area (Q_{500}) was extracted from the cyclic voltammetry curves (**Table 3**). This electrochemical parameter represents the integrated area of the cyclic voltammogram for scans taken from 0 to 0.5 V. Values of Q_{500} were in accordance with $I_{p,a}$, thus, grape pomace at early harvest, with the highest I_{pa}, showed the highest Q_{500} (0.42) followed in decreasing order by those at TH and LH (0.38 and 0.33, respectively). As shown in **Table 3**, these results were also in accordance with the total phenolic content. $E_{p,a}$ value was 0.24 V in all states of maturation.

Seeds, skins and stems from grape pomace were also measured separately at three states of maturation. **Figures 7–9** show the cyclic voltammograms for seeds, skins and stems, respectively. As can be observed, differences in voltammograms depending on the state of maturation were found. For skins and stems, $I_{p,a}$ decreased from EH to LH; however, for seeds the evolution of this parameter was different. Considering Q_{500}, differences depending on the state of maturation were found for seeds, skins and stems, and the evolution was the same as for $I_{p,a}$ (**Table 3**).

These data indicate that CV provides a reliable and good estimation of the state of maturation of winemaking by-products, because different electrochemical behaviour is shown depending on the state of maturation.

CV is also used to evaluate the differences in the antioxidant potential between grape pomaces from different variety. In our previous study [39], the electrochemical behaviour of grape pomaces

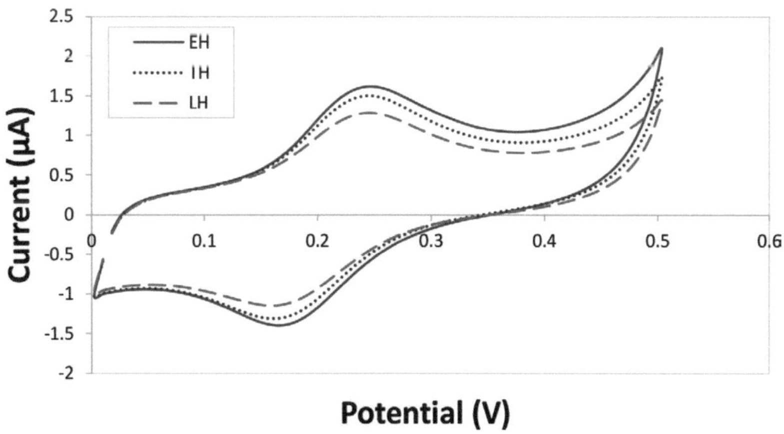

Figure 6. Cyclic voltammograms of grape pomace at different states of maturation (EH, TH, LH).

	Electrochemical parameters			
	State of maturation	$E_{p,a}$	$I_{p,a}$	Q_{500}
	EH	0.24	1.62	0.42
Pomace	TH	0.24	1.50	0.38
	LH	0.24	1.28	0.33
	EH	0.25	1.36	0.34
Seeds	TH	0.25	1.38	0.33
	LH	0.25	1.32	0.32
	EH	0.25	0.82	0.23
Skins	TH	0.25	0.78	0.22
	LH	0.25	0.75	0.21
	EH	0.25	0.79	0.23
Stems	TH	0.25	0.77	0.22
	LH	0.25	0.73	0.21

Table 3. Electrochemical parameters of anodic peak extracted from the cyclic voltammetry curves of winemaking by-products at different state of maturation.

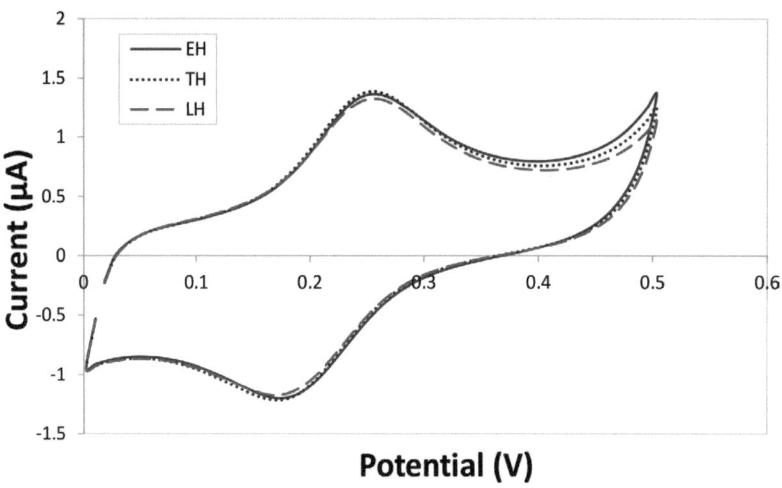

Figure 7. Cyclic voltammograms of seeds at different states of maturation (EH, TH, LH).

from nine different varieties of white grapes was studied. The cyclic voltammogram scans were made from 0 to 1 V at pH 3.6 and three different anodic peaks were observed in the voltammogram. The electrochemical parameter $I_{p,a}$ for peaks I, II and III was significantly different among varieties. Peak I was correlated mainly to phenolic acids and flavonols, peak II to flavanols and peak III to three phenolic groups. Results suggested that the electrochemical response of phenolic compounds in grape pomace extracts could be used as a measurement of the antioxidant potential.

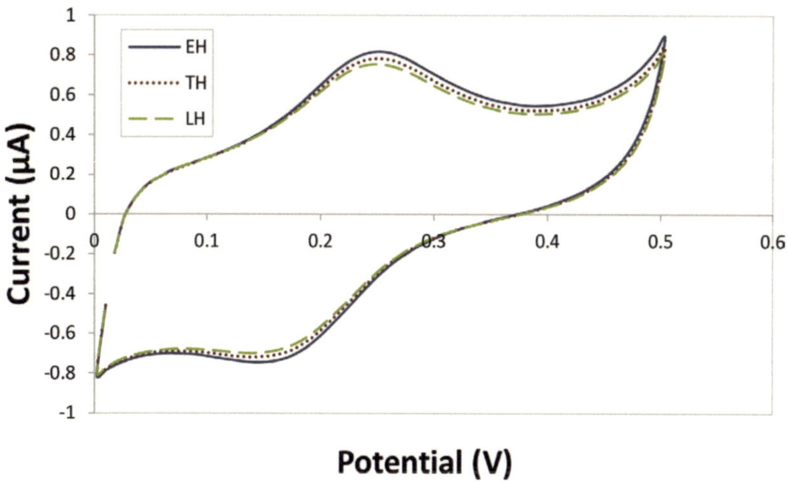

Figure 8. Cyclic voltammograms of skins at different states of maturation (EH, TH, LH).

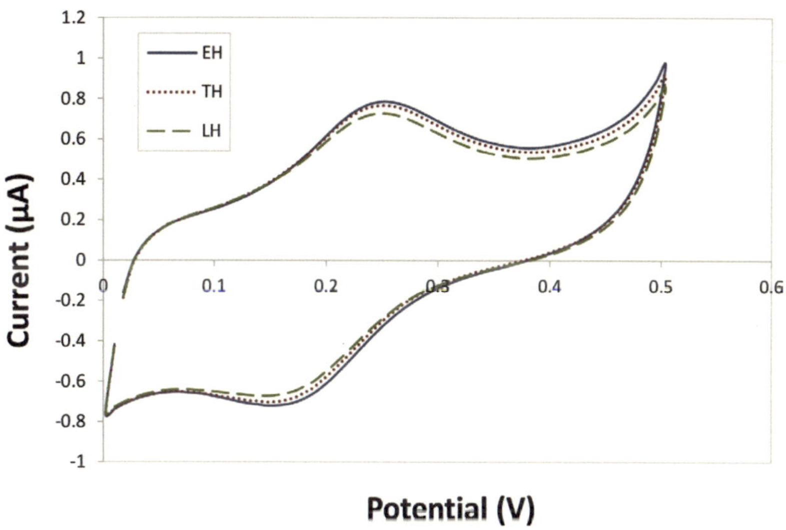

Figure 9. Cyclic voltammograms of stems at different states of maturation (EH, TH, LH).

On the other hand, CV was used as a measurement of the total antioxidant activity of different winemaking by-products: grape pomace, seeds, skins and stems [61]. The cyclic voltammograms scans were made from 0 to 0.5 V at pH 7 and the main anodic peak was examined by extracting the electrochemical parameters. Regarding $I_{p,a}$, significant differences ($p < 0.05$) were found between pomace and seeds, with higher values than skins and stems (1.44, 1.33, 0.84 and 0.73 µA, respectively). Q_{500} was used as a measure of the concentration of the total phenolic compounds and values of this parameter were significantly lower for the skins and stems (2.36 and 2.17, respectively) than for pomace and seeds (3.74 and 3.29, respectively). In this study, a principal component analysis allowed to classify between seeds, skins, stems and pomace, as a function of the electrochemical profile. Finally in this study, relationships between the voltammetric parameters (Q_{500} and $I_{p,a}$) and the results of inhibition of lipid peroxidation of winemaking by-products were explored. Results suggested that CV could be a good technique to estimate the ability of winemaking by-products to inhibit lipid peroxidation in an *in vitro* biological system.

2.2.2.2. *Wine*

Finally, CV can be used for the characterization of phenolic compounds in wine on the basis that practically all polyphenolic molecules present in wine are electrochemically active. CV was the first electrochemical method used for characterization of phenolic compounds and determination of the total phenolic content in wines [63].

CV can be utilized for direct evaluation of antioxidant activity in real samples of red wine and white wine [54, 64], and for the quantification of antioxidants on a carbon electrode in wine samples [53, 58].

In a previous work [62], we use CV to determine the electrochemical behaviour of red wines at the beginning and the end of vinification in order to study their antioxidant activity. In the cited study, three types of experimental vinification processes were performed with mixtures of Syrah grapes and addition of Pedro Ximénez seeds (simple and double dose). Grape seeds are a natural source of phenolic compounds, particularly flavanols, and their addition could improve the biological properties of wines. CV allowed establishing differences according to the area under the curve (QT, QI, QII and QIII) between control wines (without seeds) and wines with addition of seeds. Electrochemical results indicated that wines with double dose of seeds had better antioxidant activity than control wines.

3. Conclusion

Electrochemical technique, specifically cyclic voltammetry, has been used to estimate the total antioxidant potential of phenolic extract from wine products. This contribution accentuates the role of electrochemical techniques for the determination of antioxidant activity in samples.

The electrochemical behaviour of standard solutions of the main phenolic groups found in wines has been investigated and the influence of the sample matrix has been evaluated. In winemaking by-products, CV provides a reliable and good estimation of the state of maturation and the electrochemical parameters were significantly correlated to the total flavanols, flavonols and phenolic acid contents. Moreover, a good correlated was obtained between voltammetric parameters and values of lipid peroxidation inhibition *in vitro* biological system, measured by TBARS procedure. Additionally, CV allowed establishing differences in the antioxidant activity of wines with different addition of grape seeds.

Cyclic voltammetry proved to be a useful technique to estimate the antioxidant potential of wine products.

Author details

Dolores Hernanz-Vila[1], M. José Jara-Palacios[2], M. Luisa Escudero-Gilete[2] and Francisco J. Heredia[2]*

*Address all correspondence to: heredia@us.es

1 Department of Analytical Chemistry, Universidad de Sevilla, Sevilla, Spain

2 Food Color & Quality Laboratory, Department of Nutrition & Food Science, Universidad de Sevilla, Sevilla, Spain

References

[1] Horsnsey, I. The chemistry and biology winemaking. 2007. The Royal Society of Chemistry, UK. ISBN-13: 978-0-85404-266-1.

[2] Fernández de Simón B, Hernández T, Estrella I. Relationship between chemical structure and biosynthesis and accumulation of certain phenolic compounds in grape skins during ripening. Z. Lebensm. Unters. Forsch. 1992; **195**:124–128.

[3] Jackson, S. Wine science. Principles and application. 2008. Elsevier, USA. ISBN: 978-0-12-373646-8.

[4] Lorrain B, Ky L, Teissedre PL. Evolution of analysis of polyphenols from grapes, wines and extracts. Molecules. 2013; **18**:1076–1100.

[5] Kammerer DR, Carle R. Evolution of polyphenols during vinification and wine storage. Funct. Plant Sci. Biotechnol. 2009; **1**:46–59.

[6] Cheynier VF, Trousdale EK, Singleton VL, Salgues MJ, Wylde R. Characterization of 2-S-glutathioylcaftaric acid and its hydrolysis in relation to grape wines. J. Agric. Food Chem. 1986; **34**: 217–221.

[7] Sapis C, Macheix JJ, Cordonnier R. The browning capacity of grapes. I. Changes in polyphenol oxidase activities during development and maturation of the fruit. J. Agric. Food Chem. 1983; **31**:342–345.

[8] Sapis C, Macheix JJ, Cordonnier R. The browning capacity of grapes. II. Browning potential and polyphenol oxidase activities in different mature grape varieties. J. Agric. Food Chem. 1983; **34**:157–162.

[9] Oszmianski J, Ramos T, Bourzeik M. Fractionation of phenolic compounds in red wine. Am. J. Enol. Vitic. 1988; **39(3)**:259–262.

[10] Gordillo B, Rodríguez-Pulido FJ, Escudero-Gilete ML, González-Miret ML, Heredia FJ. Comprehensive colorimetric study of anthocyanic copigmentation in model solutions. Effects of pH and molar ratio. J. Agric. Food Chem. 2012; **60(11)**:2896–2905.

[11] Jara-Palacios MJ, Gordillo B, González-Miret ML, Hernanz D, Escudero-Gilete ML, Heredia FJ. Comparative study of the enological potential of different winemaking by-products: implications in the antioxidant activity and color expression of red wine anthocyanins in a model solution. J. Agric. Food Chem. 2014; **62(29)**:6975–6983.

[12] Ough CS. Substances extracted during skin contact with white musts. I. General wine composition and quality changes with contact time. Am. J. Enol. Vitic. 1969; **20**:93–100.

[13] Falqué E, Fernández E. Effects of different skin contact times on treixadura wine composition. Am. J. Enol. Vitic. 1996; **47**:309–311.

[14] Darias-Martín JJ, Rodríguez O, Díaz E, Lamuela-Raventós RM. Effect of skin contact on the antioxidant phenolics in white wine. Food Chem. 2000; **71**:483–487.

[15] Hernanz D, Recamales AF, González-Miret ML, Gómez-Mínguez MJ, Vicario IM, Heredia FJ. Phenolic composition of white wines with a prefermentative maceration at experimental and industrial-scale. J. Food Eng. 2007; **80**:327–335.

[16] Gómez-Mínguez MJ, González-Miret ML, Hernanz D, Fernández MA, Vicario IM, Heredia FJ. Effects of pre-fermentative skin contact conditions on colour and phenolic content of white wines. J. Food Eng. 2007; **32**:238–245.

[17] Teissedre PL, Frankel EN, Waterhouse AL, Peleg H, German JB. Inhibition of in vitro human LDL oxidation by phenolic antioxidants from grapes and wines. J. Sci. Food Agr. 1996; **70**:55–61.

[18] Soleas GJ, Diamandis EP, Goldberg DM. Wines as a biological fluid: history, production and role in disease prevention. J. Clin. Lab. Anal. 1996; **11**:287–317.

[19] Recamales AF, Sayago A, González-Miret ML, Hernanz D. The effect of time and storage conditions on the phenolic composition and colour of white wine. Food Res. Int. 2006; **39**:220–229.

[20] Darias-Martín J, Díaz-González D, Díaz-Romero C. Influence of two pressing processes on the quality of must in white wine production. J. Food Eng. 2004; **63**:335–340.

[21] Cheynier V, Rigaud J, Souquet JM, Barillére JM, Moutounet M. Effect of pomace contact and hyperoxidation on the phenolic composition and quality of Grenache and Chardonnay wines. Am. J. Enol. Vitic. 1989; **40**:36–42.

[22] Jara-Palacios MJ, González-Manzano S, Escudero Gilete ML, Hernanz D, Dueñas M, González-Paramás A, Heredia FJ, Santos-Buelga C. Study of zalema grape pomace: phenolic composition and biological bffects in caenorhabditis elegans. J. Agric. Food Chem. 2013; **61**:5114–5121.

[23] Jara-Palacios MJ, Hernanz D, Cifuentes-Gómez T, Escudero-Gilete ML, Heredia FJ, Spencer PE. Assessment of white grape pomace from winemaking as source of bioactive compounds, and its antiproliferative activity. Food Chem. 2015; **183**:78–82.

[24] King RE, Bomser JA, Min DB. Bioactivity of resveratrol. Comprehensive Rev. Food Sci. Food Safety. 2006; **5(3)**:65–70.

[25] Santos-Buelga C, Scalbert A. Proanthocyanidins and tannin-like compounds—nature, occurrence, dietary intake and effects on nutrition and health. J. Sci. Food Agr. 2000; **80**:1094–1117.

[26] Sies H. Oxidative stress: from basic research to clinical application. Am. J. Med. 1991; **91**:31–38.

[27] Finkel T, Holbrook NJ. Oxidants, oxidative stress and the biology of ageing. Nature. 2000; **408**:239–247.

[28] Juránek I, Bezek S. Controversy of free radical hypothesis: reactive oxygen species–cause or consequence of tissue injury? Gen Physiol. Biophys. 2005; **24**:263–278.

[29] Halliwell B. Free radicals and antioxidants – quo vadis?. Trends Pharmacol. Sci. 2011; **32**:125–130.

[30] Halliwell B. Antioxidants in human health and disease. Annu. Rev. Nut. 1996; **16**:33–50.

[31] Kohen R, Nyska A. Oxidation of biological systems: oxidative stress phenomena, antioxidants, redox reactions, and methods for their quantification. Toxicol. Pathol. 2002; **30**:620–650.

[32] Lozano C, Torres JL, Julia L, Jimenez A, Centelles JJ, Cascante M. Effect of new antioxidant cysteinyl-flavanol conjugates on skin cancer cells. FEBS Lett. 2005; **579**:4219–4225.

[33] Halliwell B. Reactive species and antioxidants. Redox biology is a fundamental theme of aerobic life. Plant. Physiol. 2006; **141**:312–322.

[34] Pham-Huy LA, He H, Pham-Huy C. Free radicals, antioxidants in disease and health. Int. J. Biomed. Sci. 2008; **4**:89–96.

[35] Apak R, Özyürek M, Güçlü K, Çapanoğlu E. Antioxidant activity/capacity measurement. 3. Reactive oxygen and nitrogen species (ROS/RNS) scavenging assays, oxidative stress biomarkers, and chromatographic/chemometric assays. J. Agric. Food Chem. 2016; **64**:1046–1070.

[36] Wang H, Cao GH, Prior RL. Total antioxidant capacity of fruits. J. Agric. Food Chem. 1996; **44**:701–705.

[37] Kähkönen MP, Hopia AI, Vuorela HJ, Rauha JP, Pihlaja K, Kujala TS, Heinonen M. Antioxidant activity of plant extracts containing phenolic compounds. J. Agric. Food Chem. 1999; **47**:3954–3962.

[38] Shui G, Leong LP. Residue from star fruit as valuable source for functional food ingredients and antioxidant nutraceuticals. Food Chem. 2006; **97**:277–284.

[39] Jara-Palacios MJ, Hernanz D, Escudero-Gilete ML, Heredia FJ. Antioxidant potential of white grape pomaces: phenolic composition and antioxidant capacity measured by spectrophotometric and cyclic voltammetry methods. Food Res. Int. 2014; **66**:150–157.

[40] Cumby N, Zhong Y, Naczk M, Shahidi F. Antioxidant activity and water-holding capacity of canola protein hydrolysates. Food Chem. 2008; **109**:144–148.

[41] Liyana-Pathirana C, Dexter J, Shahidi F. Antioxidant properties of wheat as affected by pearling. J. Agric. Food Chem. 2006; **54**:6177–6184.

[42] Shahidi F, Alasalvar C, Liyana-Pathirana CM. Antioxidant phytochemicals in hazelnut kernel (*Corylus avellana* L.) and hazelnut by-products. J. Agric. Food Chem. 2007; **55**:1212–1220.

[43] Shahidi F, Zhong Y. Measurement of antioxidant activity. J. Funct. Food. 2015; **18**:757-781.

[44] Floegel A, Kim DO, Chung SJ, Koo SI, Chun OK. Comparison of ABTS/DPPH assays to measure antioxidant capacity in popular antioxidant-rich US foods. J. Food Compos. Anal. 2011; **24**:1043–1048.

[45] Prior RL, Wu X, Schaich K. Standardized methods for the determination of antioxidant capacity and phenolics in foods and dietary supplements. J. Agric. Food Chem. 2005; **53**:4290–4303.

[46] Fontana AR, Antoniolli A, Bottini R. Grape pomace as a sustainable source of bioactive compounds: extraction, characterization, and biotechnological applications of phenolics. J. Agric. Food Chem. 2013; **61**:8987–9003.

[47] Chevion S, Roberts MA, Chevion M. The use of cyclic voltammetry for the evaluation of antioxidant capacity. Free Radical Biol. Med. 2000; **28**:860–870.

[48] Yakovleva KE, Kurzeev SA, Stepanova EV, Fedorova TV, Kuznetsov BA, Koroleva OV. Characterization of plant phenolic compounds by cyclic voltammetry. Appl. Biochem. Microbiol. 2007; **43**:661–668.

[49] Ceballos C, Fernandez H. Synthetic antioxidants determination in lard and vegetable oils by the use of voltammetric methods on disk ultramicroelectrodes. Food Res. Int. 2000; **33**:357–365.

[50] Chen J, Gorton L, Åkesson B. Electrochemical studies on antioxidants in bovine milk. Anal. Chim. Acta. 2002; **474**:137–146.

[51] Sousa WR, da Rocha C, Cardoso CL, Silva DHS, Zanoni MV. Determination of the relative contribution of phenolic antioxidants in orange juice by voltammetric methods. J. Food Composition Anal. 2004; **17**:619–633.

[52] Kilmartin PA, Zou H, Waterhouse AL. A cyclic voltammetry method suitable for characterizing antioxidant properties of wine and wine phenolics J. Agric. Food Chem. 2001; **49**:1957–1965.

[53] Kilmartin PA, Zou H, Waterhouse AL. Correlation of wine phenolic composition versus cyclic voltammetry response. Am. J. Enol. Vitic. 2002; **53**:294–302.

[54] De Beer D, Harbertson JF, Kilmartin PA, Roginsky V, Barsukova T, Adams DO, Waterhouse AL. Phenolics: a comparison of diverse analytical methods. Am. J. Enol. Vitic. 2004; **55**:389–400.

[55] Dobes J, Ondrej Z, Sochor J, Ruttkay-Nedecky B, Babula P, Beklova M, Kynicky J, Hubalek J, Klejdus B, Kizek R, Adam V. Electrochemical tools for determination of phenolic compounds in plants. A review. Int. J. Electrochi. Sci. 2013; **8**:4520–4542.

[56] Makhotkina O, Kilmartin PA. The phenolic composition of Sauvignon blanc juice profiled by cyclic voltammetry. Electrochim. Acta. 2012; **83**:188–195.

[57] Rebelo MJ, Rego R, Ferreira M, Oliveira MC. Comparative study of the antioxidant capacity and polyphenol content of Douro wines by chemical and electrochemical methods. Food Chem. 2013; **141**:566–573.

[58] Makhotkina O, Kilmartin PA. The use of cyclic voltammetry for wine analysis: determination of polyphenols and free sulfur dioxide. Anal. Chim. Acta. 2010; **668**:155–165.

[59] Petrovic SC. Correlation of perceived wine astringency to cyclic voltametric response. Am. J. Enol. Vitic. 2009; 60:373–378.

[60] Kilmartin, PA. Electrochemistry applied to the analysis of wine: a mini-review. Electrochem. Commun. 2016; **67**:39–42.

[61] Jara-Palacios MJ, Escudero-Gilete ML, Hernández-Hierro J, Heredia FJ, Hernanz D. Cyclic voltammetry to evaluate the antioxidant potential in winemaking by-products. Talanta. 2017; **165**:211–215.

[62] Jara-Palacios MJ, Hernanz Dolores, Escudero-Gilete ML, Heredia Francisco J. The use of grape seed by-products rich in flavonoids to improve the antioxidant potential of red wines. Molecules. 2016; 21:1526–1538.

[63] Šeruga, M, Novak I, Jakobek L. Determination of polyphenols content and antioxidant activity of some red wines by differential pulse voltammetry, HPLC and spectrophotometric methods. Food Chem. 2011; **124**:1208–1216.

[64] Roginsky V, De Beer D, Harbertson JF, Kilmartin PA, Barsukova T, Adams DO. The antioxidant activity of Californian red wines does not correlate with wine age. J. Sci. Food Agr. 2006; **86**:834–840.

Chapter 6

Modified Electrodes for Determining Trace Metal Ions

Pipat Chooto

Additional information is available at the end of the chapter

http://dx.doi.org/10.5772/intechopen.68193

Abstract

Due to all the advantages of low cost, speed, and simplicity, electrochemistry has always represented a perfect choice to be selected in quantitative analysis particularly in the case of metal ions but with the drawback of specificity and sensitivity. With the arrival of nanomaterials, the problem of sensitivity and limit of detection has been overcome and a great variety of applications of electrochemistry especially in trace analysis are highlighted. Layers of materials can be arranged and manipulated to make the methods more specific to targeting analytes The opportunity is there for both older and newer methods to be beneficial in a large number of applications with superb analytical performance. This knowledge of modified electrodes can inspire newer and greater innovative applications of electrochemistry with the promising extension to other areas under current interests.

Keywords: modified electrodes, ASV, nanomaterials, metal ion analysis

1. Introduction

A number of techniques have been employed for the determination of trace metal ions including atomic absorption spectrometry (AAS), inductively coupled plasma-mass spectrometry (ICP-MS), inductively coupled plasma-optical emission spectrometry (ICP-OES), and electrochemical techniques. Spectroscopic techniques are very expensive and need preconcentration as well as extraction that are time-consuming with danger of losses and contamination [1]. Electroanalytical techniques, particularly anodic stripping voltammetry (ASV), can be considered as the most powerful techniques due to their excellent detection limits, high sensitivity, capacity for multielement determination, high speed, simplicity, and relatively low cost [2] not to mention their innovative opportunities. It is important to be noted right at the very first here that voltammetry is not the only technique to be used for modified electrodes but other electrochemical techniques can be applied as well, especially potentiometry.

The selection of a proper electrode material is crucial in voltammetry. For the past six decades, mercury has been the most commonly used electrode material in various configurations for electrochemical determination of trace metal ions. Despite advantages of formation of amalgam and high overvoltage for gases among others, there have been numerous attempts to replace well-known toxic mercury with some other nontoxic or less-toxic electrode material [3]. Nowadays, numerous new electrode materials and methods have been developed, especially those concerning electrode modifications in particular with nanomaterials.

2. Background of modified electrodes

In general especially in the past, an electrode can be any electroconducting materials that were started by metals such as platinum or gold. Later, glassy carbon has been used with a number of advantages in particular ease to use and wide potential range. After that, carbon paste has been applied due to the fact that it is easy to prepare. Various substances have been mixed to attract the analytes especially metal ions to be collected at electrode surface and increase the sensitivity. With an introduction of nanomaterials and conducting polymers, for example, the surface areas for preconcentrating metal ions have been dramatically increased, making the method perfect for trace metal analysis in accordance with simplicity and low cost of electrochemical methods. Consequently, at present, there are a great number of research articles involving the development of new methods using a variety of modified electrodes to be applied with various areas as well as samples. To make this chapter simple but specific, the use of enzymes in the form of biosensors is not mentioned here. Those who are interested can obtain those specific stories in detail in a large number of available references [4]. We also have to say that modified electrodes can be used with a great variety of analytes, but metal ions are under the focus here. However, for the sake of abundant available applications and promising characteristics in adapting to metal ion analysis, the determinations of other analytes will be concisely included.

3. Types of substrate electrodes

Due to the fact that there are vast types of available and investigated substrates, the most recent and the most popular are discussed here. Other less frequently used electrodes such as carbon fiber or carbon cloth are not included. The readers are recommended to further study corresponding articles for more details.

3.1. Glassy carbon electrode (GCE)

Glassy carbon electrodes (GCEs) are prepared by means of a carefully controlled heating program of premodeled polymeric resin body in an inert atmosphere [5]. Unlike many nongraphitizing carbons, it is impermeable to gases and also resistant to acid attack. The structure of glassy carbon consists of graphite planes randomly organized in a complex topology. Glassy carbon possesses isotropic properties and does not require a particular orientation in the electrode device. The properties of carbonaceous materials significantly depend on the manufacturing processes involved. Surface treatment is usually employed to create its active and reproducible surface to enhance analytical performances. Another way is to include certain additional activation steps such as electrochemical, chemical, vacuum heat, or laser treatment.

Carbon electrodes offer a useful and environmentally friendly alternative to substitute mercury electrodes with a narrow cathodic range or noble metal surfaces with limitations in terms of reproducibility, formation of oxide layers during voltammetric procedures and relatively low cost [6]. It becomes one of the most commonly used substrates due to its wide potential window with low background and its chemical stability. Electrode modification can then be applied to improve its performance in terms of sensitivity, selectivity, and reproducibility.

3.2. Boron-doped diamond (BDD)

Boron-doped diamond (BDD) electrodes have also currently attracted much interest to be applied in a variety of areas due to their superior properties, including extreme robustness with a low level of background interference, less adsorption of polar molecules, and attractively wider potential window in aqueous media [7, 8]. It has been used to quantify manganese in tea [9] as well as lead in tap water [10] and river sediment. Anodic stripping voltammetry BDD has been proved to possess outstanding features [11] to determine silver [12] and simultaneous detection of lead and copper [13].

3.3. Fluorine-doped tin oxide (FTO)

Fluorine-doped tin oxide (FTO) has been applied continuously as a substrate with outstanding features of simplicity in layer-by-layer (LbL) fabrication and its compatibility with extensive building blocks including dyes, biomolecules, nanomaterials, and polymers [14]. In spite of the fact that it has been reported to be successfully applied in the analysis of biosubstances particularly DNA, it is also mentioned here in light of making its promising way to metal ion analysis.

3.4. Screen-printed electrode (SPE)

There are numerous possibilities to choose from for screen-printed electrode (SPE). The most popular material is still carbon. SPE has advantages of small size, low cost, simplicity as well as smaller amount of sample and waste. The problem of lower sensitivity can be solved by electrode modification, which also highlights its applications in a larger number of areas [15].

3.5. Carbon paste electrode (CPE)

Carbon paste is still widely used throughout the development of modified electrodes with certain reasons including superb quality of carbon as an electrode, low cost, and its simplicity [16]. With clever design, additional benefits can be reached including stability, reproducibility, and fast response time. This material has been found to be useful for the determination of both compounds and metal ions.

3.6. Silica

Silica, in particular mesoporous silica, has been increasingly used in modified electrode with features of inertness, high surface area, moderate cost, availability, and compatibility of being anchored by various materials. It has been reported to be useful in the analysis of both biomolecules and metal ions [17].

4. Types of modified electrodes

A number of materials have been investigated to be used in preconcentrating metal ions as well as other substances and make electrochemistry unique and highlighted in the worlds of analytical chemistry and beyond. Thanks to the developments and arrivals of nanomaterials, the most widely used especially at the very beginning is metal nanoparticles such as silver or gold to increase the surface areas and in turn the sites for metal ions to deposit. Both conducting and nonconducting polymers have been used for a long time in modifying electrode surface to have more capabilities in supporting metal ions. Mesoporous silica with the advantage of surface areas as well has been used in the determinations of a number of metal ions. Another example of a neutral substance with greater surface areas in collecting metal ions is chitosan, a substance from shrimp. Currently, it is certain that the opportunity is there that a large number of substances are under investigations or even await the discovery. Finally, the combinations of a variety of materials have also been proved to be useful in further receiving the metals ions to a greater extent. The electrodes modified by aforementioned materials are then applied in stripping voltammetry, parameters are optimized, and then the methods are used with real samples. Normally, the results are compared with standard methods or the standard materials are used for verification. A number of spectroscopic and electrochemical methods can also be used to provide additional details of the analysis. At present, a very large number of research articles focus on the applications of modified electrodes in many areas especially in the analysis of a great variety of substances, in particular, metal ions. Also, a number of materials have been investigated in the form of layers and sublayers as well as specific pores as a specific substrate for particular analytes, hence, the new term of "molecular imprinted," which makes the method extremely specific.

The following materials that have been used in electrode modifications are not arranged with the criteria of the time of development. Rather, it is presented in the order of simplicity.

4.1. Unmodified electrode

With a superb characteristic of specific electrode such as screen-printed carbon electrode, metal ion can still be determined at trace level by in a very normal way [18].

4.2. Graphene

Graphene is an allotrope of carbon in the form of a two-dimensional, atomic-scale and hexagonal lattice in which one atom forms each vertex. It is composed of a single layer of sp2 carbon in two dimensions. It is the basic structural element of other allotropes, including graphite, charcoal, carbon nanotubes (CNTs), and fullerenes. Graphene has a great variety of unusual beneficial properties including strength, heat and electricity conductivity, transparency, magnetic properties, and low cost [19].

4.3. Graphene oxide

Graphene can be prepared in a modified way to obtain different and beneficial properties in new forms including thermally reduced graphene, partially reduced graphene, or even electrochemically reduced graphene (ErGO). Normally, this is the arrangement of oxygen in the structure, hence the name graphene oxide that is really helpful in collecting metal ions and providing better selectivity, resolution, as well as precision. With the addition of other substance that can form the bond via conjugation with graphene, electrocatalization as well as electroluminescence (ECL) can be facilitated. This modified graphene derivatives can be use satisfactorily in both waste water treatment via adsorption [20] as well as analysis in only one step [21] in addition to the development of new batteries [21, 22] and improvement of antibacterial properties [23].

4.4. Metals

Metal and metal alloys can also be used in the analysis of different species such as nitrite but the applications for metal ions are focused here. Moreover, as a typical case,only metal that can satisfactorily substitute mercury namely bismuth is emphasized.

In 2000, a new type of electrode called bismuth film electrode (BiFE) consisting of a thin film of bismuth deposited on a carbon substrate has been proposed as an alternative to mercury electrodes in ASV [24]. The main advantage of electrochemical properties of bismuth film electrodes in comparison with mercury film electrodes (MFEs) is that Bi is more environmentally friendly with less toxicity in addition to simple preparation, high sensitivity, well-defined and separated stripping signals, and insensitivity to dissolved oxygen (which is an essential property for on-site monitoring). The superior stripping performances of bismuth-based electrodes derive from their ability to form "fused" alloys with other metals similar to mercury [24].

There are three common ways to generate a bismuth film including (i) by preplating it from an acidic solution which is called an ex situ preparation, (ii) by codeposition with the analyte which is commonly known as an in situ setup and (iii) by electrode modification of a film, such as Bi_2O_3(s) or BiF_3, to generate the Bi(s) coating [25]. Ex situ plating was found to be easier to manage because the conditions can be different from analytical or stripping conditions, and there are no interferents in depositing; however, it is more susceptible to the change of electrode surface during electrode transfer and more steps make the method take longer time. Another advantage of ex situ methods is that the electrode can be regenerated at any time. Also, the potential can be better controlled due to the fact that, for in situ preparation, the stripping of bismuth needs to be performed at the potential more positive than bismuth oxidation and after that bismuth is replated [26].

4.5. Metal complexes

A number of metal complexes have been immobilized on the substrate to attract or react with other substances. Due to the fact that it already contains metals, this type of modification substance is normally used for the determination of organic and inorganic compounds especially via electrocatalysis [27]. Cobalt phthalocyanin has been widely and continuously investigated and applied for the analysis of ascorbic, diethyl stilbestol, and acetaminophen [28]. Manganese porphyrins have been extensively investigated [29]. As a matter of fact, porphyrins themselves can accommodate metal ions really well and, with the increase of surface areas, should be able to be used in the analysis of metal ions [30].

4.6. Metal nanoparticles

There was a wonderful review for metal nanoparticles for the determination of arsenic, chromium, lead, cadmium, and antimony [31]. Mixing metal nanoparticles with a wide range of compounds can allow the analytical performances of the methodology to be greatly improved in various aspects especially sensitivity due to larger amount of analytes collected.

4.7. Metal compound nanoparticles

Due to the fact that there are a great variety of metal compound nanoparticles that have been used in metal ion analysis especially recently [32], only modified magnetic iron oxide nanoparticles (M-MIONPs) for mercury determination are mentioned here as an example.

It is well known that mercury in the lowest levels of concentrations is dangerous for human health due to its bioaccumulation in body and toxicity. Modified magnetic iron oxide nanoparticles (M-MIONPs) with 2-mercaptobenzothiazole (MBT) was found to be able to absorb mercury (II) ion satisfactorily from polluted surface water with advantages of speed, cost-effectiveness, simplicity, capability, ease of preparation, and safety [33]. Modification by 2-mercaptobanzothiazole could increase absorption percentage up to 98.6% compared with 43.47% for magnetic iron oxide nanoparticles (MIONPs) alone. Salt concentrations and pH were found to have no profound effect on mercury ion accumulation with high loading capacity of 590 μg/g. This proves that the capability of metal compound nanoparticles in attracting analytes can be greatly improved by combining them with additional compounds.

4.8. Organic compounds

Organic compounds that can be used normally or after polymerization are provided in the topic of polymers. All kinds of organic compounds that can attract metal ions can be used well in metal ion determination. The stronger bond obtained from the compounds, the better they can be applied in accumulating metal ions. Ketones and quinones form another group of interest with specific interaction with certain metal ions [34]. Additionally, all organic compounds can be made nanostructured by mounting in a multilayer form on substrate electrode. A few popular compounds are exemplified as follows.

4.8.1. Crown ether

Crown ether is a macrocyclic compound with a pore of specific size to accommodate metal ions. With derivation, its selectivity can be greatly increased. This characteristic combined with different potential of stripping makes the methodology suitable for simultaneous determination of metal ions which can face or cause interferences in other techniques [35].

Strategies can also be designed to let the compounds to form self-assembled monolayers (SAM) on metal electrodes or to be immobilized on other monolayers [36, 37].

4.8.2. *Schiff bases*

Schiff bases are defined as the substances that contain the C=N moiety. With their specific capability in forming complexes with metal ions, Schiff bases can help increase the quantity of analytes on the electrode surface. Typical examples are potentiometric determination of Co(II) [38] and cyclic voltammetric analysis of Al(III) [39].

4.9. Polymers

Two cases of 2-mercaptobenzothiazole and diazonium are stated here for the vision about the applications with the use of materials in this group that can be in both monomeric and polymeric forms. Moreover, certain polymers can also be used for the purpose of molecular imprint [40].

4.9.1. *2-mercaptobenzothiazole*

2-mercaptobenzothiazole (MBT) has been found in both monomer and polymer forms with the capabilities of collecting metal ions. Modification of nano-TiO_2 modified with 2-mercaptobenzothiazole (MBT) was found to be capable of collecting metal ions including Cd(II), Cu(II), and Pb(II) followed by elution with nitric acid and analysis by flame AAS [41]. Adsorption process as well as analytical conditions was optimized to obtain the dynamic range in ng/ml of 0–25.0 for Cd, 0.2–20.0 for Cu and 3.0–70.0 for Pb. The method was applied to the determination of Cd(II), Cu(II), and Pb(II) in water and ore samples. Obviously, this can also be applied to the analysis by electrochemistry without any need for elution. As a matter of fact, this is the topic under investigations of our group at present.

Poly(2-mercaptobenzothiazole) (PMBT) modified glassy carbon electrode has been fabricated and employed for the determination of specific organic compounds namely dopamine (DA), uric acid (UA), and nitrite (NO_2^-) in pH 6 phosphate buffer [42]. PMBT was found to catalyze oxidation of the compounds and shift the potentials to more negative which in turn resulted in well-defined and well-separated differential pulse (DP) peaks and made them possible to be simultaneously analyzed. SEM also revealed that continuous PMBT was formed with nano-scaled particles of 15–25 nm diameters. With optimized conditions, dynamic linear range in μmol/l was found to be 0.8–45 for DA, 0–165 for UA, and 60–1000 for NO_2^- with excellent linearity and submicromolar detection limits. Moreover, using standard addition, the methodology could be applied well with the real samples of urine and serum. Once again, due to the fact that the compound can react with metal ions well, this could shed some lights on simultaneous analysis of metal ions as well.

4.9.2. *Diazonium*

The modification through the electrochemical or chemical reduction of aromatic diazonium derivatives has been extensively investigated on a variety of carbon substrate including glassy carbon [43, 44], graphite [45], graphene [46], and carbon nanotube [47]. It has been proved to immobilize a great variety of functional groups onto carbon materials with simplicity and versatility to be used in metal analysis in a number of areas. Another advantage is long-term stability both in air and organic solvents. The high stability of the diazonium-modified electrodes and the versatility of the diazonium modification method are particularly attractive for stripping analysis. Carbon modified by the reduction of aromatic diazonium derivatives was first used as an electrode for electrochemical stripping analysis of heavy metals [44]. Diazobenzoic acid was reduced on GCE to obtain benzoic acid modified GCE to simultaneously analyze Cd^{2+} and Pb^{2+}. The sensitivity of stripping peaks for both metals was increased up to six times with satisfactory analytical performances including 0.5–50 μg/l linear range, submicrogram per liter detection limits, and superbly low relative standard deviation especially for Cd^{2+}. The method was successfully used in determining the metals in sewage samples. The detection of Cd^{2+} by ASV on BDD electrode based on simple and selective electrochemical reduction of Cd^{2+} on diazonium-modified BDD electrode has been developed with analytical performance interference study as well as verification by analyzing standard material. The method was then applied to the analysis of Cd in tap water [43].

4.10. Chitosans (natural polymers)

Chitosan (CTS), poly-[1,4]-N-D-glucosamine, is one of the most abundant natural polymers. Its pKa is about 6.5; therefore, at lower pH solutions (>pKa), its primary amines are protonated, making it a cationic polyelectrolyte that is soluble in aqueous solution. At higher pH (>pKa), these amines are deprotonated which, in turn, makes chitosan neutral and insoluble [48]. The reasons that chitosan can be applied well in the analysis of drug substances, environment pollutants, industrial materials, and food compounds are that they can form the film well and attach strongly to the surfaces. They are also hydrophilic, compatible with biological substances, mechanical resistant, and capable to be further modified [49].

4.11. Clay

It has long been known that cationic metals can be strongly absorbed on clay materials with negative charge. A large number of scientists especially in the areas of environments have extensively studied the adsorption of metal ions on the clay particles. This characteristic also benefits the determination as well as elimination of metal ions [50, 51].

4.12. Mesoporous silica

Mesoporous materials are described as materials whose pore diameters lie in the range between 2 and 50 nm [52]. These materials are in focus due to the fact that they have abundant surface areas, they can absorb metal ion very fast, and their pore size as well as pore arrangement can be well-controlled. Moreover, they can be chemically modified with other functional groups to be able to better attract large variety of metal ions for the purpose of simultaneous analysis and removal for various samples [53].

4.13. Charcoal

Due to the fact that different kinds of charcoal can specifically adsorb metal ions on their surface [54, 55], they should work well in collecting metal ions. The increase of both surface areas and specificity from modifications can facilitate better analytical performances. Even though there have not yet been recent reports about their applications in metal ion analysis, the opportunity is there to apply charcoals onto substrates as a new methodology to reach the objective of using readily obtained and low-cost materials in both analysis and removal of metal ions.

4.14. Carbon nanotube

Carbon nanotubes are tube-form materials with the diameter at nanometer level discovered by a Japanese scientist, Sumio Iijima, in 1991. They can be classified into single-walled (SWCNT) and multiwalled (MWCNT) with different properties especially in terms of metallic and magnetic behavior. They can be prepared by chemical vapor deposition (CVD), arc discharge, or laser vaporization. They can be applied in a large number of areas especially modified electrodes. Carbon nanotubes can be mounted either alone or mixed with other materials on any substrate electrode but preferably GCE. MWCNT is normally more satisfactory due to its advantages of highly ordered structure, light weight strength as well as thermal and electrical conductivity. In particular, the multi-walled have been extensively used in the determination of organic compounds [56] or metal ions either by electrochemistry [57] or spectroscopy [58, 59]. Their advantages in analysis mainly derive from the capabilities to adsorb metal ions [60]. This property makes it suitable to be applied in the areas of energy [61]. Furthermore, with large surface areas of carbon nanotubes, a number of substances can be mounted on them either single layer or multilayer to increase the capability to preconcentrate metal ions before their determinations [62].

4.15. Mixed or multilayered modification

Mixed materials can be used to determine both organic and inorganic substances including metal ions with the only reason of selectivity improvement. Despite of the fact that there are increasing methods to determine compounds such as H_2O_2 or glycerol, the combination of modified materials has been proved to facilitate the determination of trace metals. The good example is the use of bismuth, polystyrene sulfonate (PSS), and carbon nanopowder (CnP) in the determination of cadmium and lead [63]. This group can be further researched with the keyword "nanocomposites" [23, 64].

4.16. Biomolecules

Certain biomolecules including DNA, peptides, algae, and cell among numerous others can be used to determine specific metal ions. However, the experimental procedures can be much more complicated and difficult. The readers are recommended to obtain more information from an available review [65].

5. Roles of trace metal ions

Heavy metal contaminations have become one of the environmental issues of global concern due to the serious harm to human health. They have been main contribution for environmental problems caused by their ecological toxicity in a number of areas worldwide. Heavy metals and their products have been extensively distributed in natural surroundings, and they continued their cycles in accumulating in living organisms before passing on to human.

Among those not easily removed from the environment are cadmium, mercury, copper, lead, silver, zinc, and arsenic [41]. Lead and cadmium are responsible for the damage of kidney and nervous as well as circulation systems [66]. Lead particularly has the greatest effects on children due to the fact that it causes irreversible neurological disorders. The limits of lead and cadmium in drinking water set in the USA are 0.015 and 0.005 mg/l respectively [67]. Therefore, control and accurate determination of trace metals in environment is of paramount importance.

6. Stripping techniques for metal ion determination

For voltammetry, stripping techniques are the most widely used in metal ion analysis [2, 3] and normally the main objective of developing new ASV methodology for is to improve the analytical performances in determining trace metal ions including higher reproducibility, higher sensitivity, more convenience, better speed, lower cost, and environmentally friendlier conditions. The methods are optimized as well as standardized and then applied to the analysis of a great variety of real samples. Their brief practical aspects are presented as follows.

7. Optimizations of stripping voltammetry

After the modified electrode of interest is fabricated and its characteristics such as wettability are clearly defined, involving parameters are optimized such as electrolyte and electrolyte concentrations, pH and buffer to use, concentration of modifying agent and involving materials, deposition potential, deposition time, scan rate, and interferences. The optimized method is then applied with standards to obtain analytical performances followed by methods validations. Finally, real samples can be analyzed in comparison with other standard methods.

8. Comparison of voltammetry with other methods

The comparison of voltammetry with normal electrode has been comprehensively discussed, especially for the speciation of arsenic [68]. Spectroscopic methods can provide the best limit of detection (LOD) but with high cost. With higher LOD, voltammetry is a better choice. Due to much greater sensitivity achieved by using modified electrodes, previous obstacles can be overcome and makes a large number of methods in the past applicable to real sample analysis by electrochemistry.

9. Study of metal ligand interaction and surface

Once practical approaches have been clearly proved to be applicable, the next important step is delving into involving interactions in order to lay the brick for future development of modifying materials as well as metal species to be determined. Methods such as X-ray crystallography, cyclic voltammetry (CV), Electrochemical Impedance Spectroscopy (EIS), and quantum calculations can be helpful in understanding collecting interaction and bond formation between metal ions and coordinating atoms [69, 70].

In addition, normally surface method such as Scanning Electron Microscopy (SEM) as well as Transmission Electron Microscopy (TEM) can be employed to follow the change of the surface during modifications and EIS has also proved to be helpful in checking the conductivity of electrode materials [71].

10. Comparison of analytical performances for individual analyte

To picture the figures of merit and analytical performances and to compare a wide range of modified electrodes, a number of investigations have been summarized in **Tables 1–5**). The decision has been made to arrange the research items with the criteria of individual analyte with a wide range of publication periods to suit specific areas of researchers and to

Entry	Modified electrode	Methods	Ion/compound	Linear range (mol/l)	LD (nmol/l)	Ref
1	Fe_3O_4 NPs-CS[a]/GCE	DPV[j]	Bisphenol A (BPA)	0.05–30.0	8.0	[72]
2	CMK-3/nano-CILPE[b]	LSV[k]	Bisphenol A (BPA)	0.2–150	50.0	[73]
3	Fe_3O_4 NPs-CB[c]/GCE	DPV	Bisphenol A (BPA)	0.0001–50.0	0.031	[74]
4	Au NPs/SGNF[d]/GCE	LSV	Bisphenol A (BPA)	0.08–250.0	35.0	[75]
5	Au NPs-GR[e]/GCE	DPV	Bisphenol A (BPA)	0.0001–100	50.0	[76]
6	Fe_3O_4 NPs-PANAM[f]/GCE	AMP[l]	Bisphenol A (BPA)	0.01–3.07	5.0	[77]
7	RGO[g]/CNT[h]/Au NPs/SPE[i]	DPV	Bisphenol A (BPA)	0.00145–1.49	0.8	[78]

[a]CS: chitosan.
[b]CMK-3/nano-CILPE: ordered mesoporous carbon modified nano-carbon ionic liquid paste electrode.
[c]CB: carbon black.
[d]SGNF: stacked graphene nanofibers.
[e]Au NPs-GR: gold nanoparticles dotted graphene.
[f]PANAM: poly(amidoamine).
[g]RGO: reduced graphene oxide.
[h]CNT: carbon nanotubes.
[i]SPE: screen-printed electrode.
[j]DPV: Differential Pulse Voltammetry
[k]LSV: Linear Scan Voltammetry
[l]AMP: Amperometry

Table 1. Analytical performances of various modified electrodes for BPA determination.

Entry	Modified electrode	Methods	Ion/compound	Linear range (μg/l)	LD (μg/l)	Ref
1	CB-15-crown-5[a]/GCE	DPASV[j]	Pb/Cd	10.9–186.5/15.7–191.1	3.3/4.7	[35]
2	BiOCl[b]/MWCNT[c]/GCE	SWASV[k]	Pb/Cd	5–50/5–50	0.57/1.2	[79]
3	L-cys[d]/GR[e]-CS/GCE	DPASV	Pb/Cd	1.04–62.1/0.56–67.2	0.12/0.45	[80]
4	MWCNT/poly(PCV)[f]/GCE	DPASV	Pb/Cd	1.0–200.0/1.0–300.0	0.4/0.2	[81]
5	Bi-D24C8[g]/Nafion SPCE	SWASV	Pb/Cd	0.5–60/0.5–60	0.11/0.27	[2]
6	Bi/poly(p-ABSA)/GCE	DPASV	Pb/Cd	1.0–130/1.0–110.0	0.8/0.63	[82]
7	Bi-xerogel/Nafion/GCE	SWASV	Pb/Cd	1.04–20.72/0.56–11.24	1.3/0.37	[83]
8	Bi/CNT/SPE	SWASV	Pb/Cd	2–100/2–100	0.2/0.8	[84]
9	Bi_2O_3/GCE[h]	SWASV	Pb/Cd	2–250/1–150	0.26/0.52	[85]
10	BiF_4/CPE[i]	SWASV	Pb/Cd	20–100/20–100	9.8/1.2	[86]

[a]CB-15-crown-5, 4-carbox-ybenzo-15-crown-5.
[b]BioCl, bismuth-oxychloride.
[c]MWCNT, multi-walled carbon nanotube.
[d]L-cys, L-cysteine.
[e]GR, graphene.
[f]poly(PCV), poly(pyrocatecholviolet).
[g]D24C8, dibenzo-24-crown-8.
[h]Bi_2O_3/GCE, graphite-composite electrodes bulk-modified with Bi_2O_3.
[i]BiF_4/CPE, ammonium tetrafluorobismuthate bulk-modified carbon paste electrode.
[j]DPASV: Differential Pulse Anodic Stripping Voltammetry.
[k]SWASV: Square Wave Anodic Stripping Voltammetry.

Table 2 Analytical performances of various modified electrodes for Pd and Cd simultaneous determination.

Entry	Modified electrode	Methods	Ion/compound	Linear range (μM)	LD (μM)	References
1	Hb[a] microbelt/GCE	CV	H_2O_2	10–230	0.61	[87]
2	HRP[b]/DNA[c]-Ag/GCE	CV	H_2O_2	7.0–7.8	2	[88]
3	Cobalt oxide NPs/GCE	CV	H_2O_2	1–1000	0.6	[89]
4	Cyt c[d]/Ag NPs/GCE	CV	H_2O_2	8.5–130	9.8	[90]

http://dx.doi.org/10.5772/intechopen.68193

Entry	Modified electrode	Methods	Ion/compound	Linear range (μM)	LD (μM)	References
5	Mb[e](Hb, HRP)/SWCNT-CTAB[f]/GCE	CV[i]	H_2O_2	24.2–1670	8.07	[91]
6	Hb/undoped nanocrystalline diamond/GCE	CV	H_2O_2	2–25	0.4	[92]
7	Hb/PAN-co-PAA[g]/GCE	CV	H_2O_2	–	4.5	[93]
8	Hb/chitosan and nano$CaCO_3$/GCE	CV	H_2O_2	–	8.3	[94]
9	Hb/nano-gold/ITO[h]	CV	H_2O_2	10–700	4.5	[95]
10	Hb/nano-Ag sol-gel/GCE	CV	H_2O_2	1–250	0.1	[96]
11	Hb/nano-Ag-chitosan/GCE	CV	H_2O_2	0.75–216	0.2	[97]

[a]Hb: Hemoglobin.
[b]HRP: Horseradish peroxidase.
[c]DNA: Deoxyribonucleic acid.
[d]Cyt c: Cytochrome c.
[e]Mb: Myoglobin.
[f]SWCNT-CTAB: Single walled carbon nanotubes-cetylramethylammonium bromide.
[g]PAN-co-PAA: poly(acrylonitrile-co-acrylic acid).
[h]ITO: Indium tin oxide.
[i]CV: Cyclic voltammetry.

Table 3. Analytical performances of various modified electrodes for H_{22} determination.

Entry	Modified electrode	Methods	Ion/compound	Linear range (nM)	LD (nM)	References
1	NN[a]HMDE[b]	CSV[f]	Iron	–	0.08	[98]
2	DHN[c]/HMDE	CSV	Iron	–	0.005	[99]
3	DHN[d](mercury coated, gold, micro-wire electrode)	CSV	Iron	–	0.1	[100]
4	5 Br-PADAP[d]HDME	DLSAV[g]	Iron	0.25–100	–	[101]
5	-(IL-rGO/AuNDs[e]/Nafion/GCE)	SWV[h]	Iron	300–100,000	35	[102]

[a]NN: 1-nitroso-2-naphthol.
[b]DHN: 2,3-dihydroxynaphthalene.
[c]HDME: Hanging mercury drop electrode.
[d]5-Br-PADAP: 2-(5′-bromo-2′-pyridylazo)-5-diethylaminophenol
[e]IL-rGO/Au NDs: ionic liquid-reduced graphene oxide supported gold nanodendrites.
[f]CSV: Cathodic stripping voltammetry.
[g]DLSAV: derivative linear sweep adsorption voltammetry
[h]SWV: Square wave voltammetry.

Table 4. Analytical performances of various modified electrodes for iron determination.

Entry	Modified electrode	Methods	Ion/compound	Linear range (μg/l)	LD (μg/l)	References
1	HMDE[a]	DPASV	Se (IV)	1.2–75	–	[103]
2	BiFE[b]	DPASV	Se (IV)	2.0–30	0.1	[104]
3	AuE[c] modified with poly 3,3′-diaminobenzidine 4HCl-Nafion	DPASV	Se (IV)	0.4–158	0.06	[105]
4	Screen printed graphite electrode	DPASV	Se (IV)	10–1000	4.9	[106]
5	Au NPs/BDD	DPASV	Se (IV)	10–100	–	[107]
6	Poly(3,3′- diaminobenzidine) film/AuE	DPASV	Se (IV)	7.9–79	0.78	[108]
7	Renewable silver annular band working electrode	DPASV	Se (IV)	1.0–10	0.15	[109]
8	AuNPs/E[d](GCE)	SWASV	Se (IV)	15–55	0.12	[110]

[a]HDME: Hanging Mercury Drop Electrode.
[b]BiFE: Bismuth film electrode.
[c]AuE: Gold electrode.
[d]E: Electrochemically prepared.

Table 5. Analytical performances of various modified electrodes for Se determination.

shed light on their upcoming research. Even though the focus is on metal ions, bisphenol A and hydrogen peroxide have been used as a model for the applications of modified electrodes in analyzing other compounds. Despite of the fact that two units of concentration are expressed, the advantage of modified electrodes in moving up to better sensitivity and specificity as well as their more useful and more innovative applications in the near future can be clearly seen.

11. Future trends

Electrochemistry has been used and studied for a long time, which lays great fundamentals for the development of newer electrochemical techniques. Valuable previous discoveries await their improvements by using modified electrodes. Innovations are underway to analyze metal ions with greater analytical performances as well as to suit simultaneous determinations. New compounds can be investigated and mixed or immobilized to increase the surface areas and serve species imprints which in turn require deeper investigations for the attractions and interactions between modified substrate and analytes. Modified electrodes should also work well with spectroscopic, separation, and other methods in a variety of ways. They have already been proved to facilitate reactions for energy research [111]. The new thing that has not been considered is the use of modified electrodes in organic synthesis to make it more specific [112]. Moreover, modified electrode has already found its ways in spectroelectrochemical investigation [113]. Finally, new theoretical explanations can be adapted for better understanding and applications, which would be the stepping stones for more and greater inventions in the future.

12. Conclusions

Modified electrodes have been proved to be effective in the determination of a number of metals ions. With the speed, simplicity, and sensitivity of stripping voltammetry, the methods can be successfully applied to their analysis at trace level. Mixtures of various compounds await the art to manifest them in increasing the sensitivity for monitoring the concentrations of important metal ions. Additionally, the discovery of new nanomaterials would give stripping voltammetry a bright future. Furthermore, new electrochemical techniques such as EIS would assist the applications of modern modified electrodes in a great variety of areas. It is hoped that this article fires up researchers as well as opens up new opportunities in initiating and conducting new electrochemical research to be universally applicable in vast areas.

Author details

Pipat Chooto

Address all correspondence to: pipat.c@psu.ac.th

Analytical Chemistry Division, Department of Chemistry, Faculty of Science, Prince of Songkla University, Hatyai, Songkhla, Thailand

References

[1] Somer G, Çalscan AC, Sendil O. A new and simple procedure for the trace determination of mercury using differential pulse polargraphy and application to a salt lake sample. Turkish Journal of Chemistry. 2015;**39**:639–647

[2] Kaewkim K, Chuanuwatanakul S, Chailapakul O, Motomizu S. Determination of lead and cadmium in rice samples by sequential injection/anodic stripping voltammetry using a bismuth film/crown ether/Nafion modified screen-printed carbon electrode. Food Control. 2013;**31**:14–21

[3] Dal Borgo DS, Jovanovski V, Pihlar B, Hocevar SB. Operation of bismuth film electrode in more acidic medium. Electrochimica Acta. 2015;**155**:196–200

[4] Turdean GL. Design and development of biosensors for the detection of heavy metal toxicity. International Journal of Electrochemistry. 2011;**2011**:1–15. DOI: 10.4061/2011/343125

[5] Silwana B. Heavy and Precious Metal Toxicity Evaluation Using a Horseradish Peroxidase Immobilised Biosensor. Degree of Master, South Africa: Department of Chemistry, Faculty of Science, University of the Western Cape; 2012

[6] Thalita RS, Priscila C, Éder TGC. Simultaneous voltammetric determination of Zn(II), Pb(II), Cu(II), and Hg(II) in ethanol fuel using an organofunctionalized modified graphite-polyurethane composite disposable screen-printed device. Electroanalysis. 2014;**26**:2664–2676

[7] McGaw EA, Swain GM. A comparison of boron-doped diamond thin-film and Hg-coated glassy carbon electrodes for anodic stripping voltammetric determination of heavy metal ions in aqueous media. Analytica Chimica Acta. 2006;**575**:180–189

[8] Fierro S, Watanabe T, Akai K, Einagaa Y. Highly sensitive detection of Cr^{6+} on boron doped diamond electrodes. Electrochimica Acta. 2012;**82**:9–11

[9] Saterlay AJ, Foord JS, Compton RG. Sono-cathodic stripping voltammetry of manganese at a polished boron-doped diamond electrode: Application to the determination of manganese in instant tea. Analyst. 1999;**124**:1791–1796

[10] Chooto P, Wararatananurak P, Innuphat C. Determination of trace levels of Pb(II) in tap water by anodic stripping voltammetry with boron-doped diamond electrode. ScienceAsia. 2010;**36**:150–156

[11] Compton RG, Coles BA, Holt K, Foord JS, Markenk F, Tsai YC. Microwave-enhanced anodic stripping detection of lead in a river sediment sample. A mercury-free procedure employing a boron-doped diamond electrode. Electroanalysis. 2001;**13**:831–835

[12] Saterlay AJ, Marken F, Foord JS, Compton RG. Sonoelectrochemical investigation of silver analysis at a highly boron-doped diamond electrode. Talanta. 2000;**53**:403–415

[13] Prado C, Wilkins SJ, Marken F, Compton RG. Simultaneous electrochemical detection and determination of lead and copper at boron-doped diamond film electrodes. Electroanalysis. 2002;**14**:262–272

[14] Manzanares-Palenzuela CL, Fernandes EGR, Lobo-Castañón MJ, López-Ruiz B, Zucolotto V. Impedance sensing of DNA hybridization onto nanostructured phthalocyanine-modified electrodes. Electrochimica Acta. 2016;**221**:86–95

[15] Saengsookwaow C, Rangkupan R, Chailapakul, O, Rodthongkum N. Nitrogen-doped grapheme-polyvinylpyrrolidone/gold nanoparticles modified electrode as a novel hydrazine sensor. Sensor and Actuators B: Chemical. 2016;**227**:524–532

[16] Cazula BB, Lazarin AM. Development of chemically modified carbon paste electrodes with transition metal complexes anchored on silica gel. Materials Chemistry and Physics. 2017;**186**:470–477

[17] Jal PK, Patel S, Mishara BK. Chemical modification of silica surface by immobilization of functional groups for extractive concentration of metal ions. Talanta. 2004;**62**:1005–1028

[18] Velmurugun M, Thirumalraj B, Chen S-M, Al-Hemaid FMA, Ali MA, Elshikh MS. Development of electrochemical sensor for the determination of palladium ions (Pd^{2+}) using flexible screen printed un-modified carbon electrode. Journal of Colloid and Interface Science. 2017;**485**:123–128

[19] Zhu H, Xu Y, Liu A, Kong N, Shan F, Yang W, Barrow CJ, Liu J. Graphene nanodots-encaged porous gold electrode fabricated via ion beam sputtering deposition for electrochemical analysis of heavy metal ions. Sensors and Actuators B: Chemical. 2015;**206**:592–600

[20] Peng W, Li H, Liu Y, Song S. A review on heavy metal ions adsorption from water by graphene oxide and its composites. Journal of Molecular Liquids. 2017;**230**:496–504

[21] Thiruppathi AR, Sidhureddy B, Keeler W, Chen A. Facile one-pot synthesis of fluorinated graphene oxide for electrochemical sensing of heavy metal ions. Electrochemistry Communications. 2017;**76**:42–46

[22] Wu Y, Zhan L, Huang K, Wang H, Yu H, Wang S, Peng F, Lai C. Iron based dual-metal oxides on graphene for lithium-ion batteries anode: Effects of composition and morphology. Journal of Alloys and Compounds. 2016;**684**:47–54

[23] Sahraei R, Ghaemy M. Synthesis of modified gum tragacanth/graphene oxide composite hydrogel for heavy metal ions removal and preparation of silver nanocomposite for antibacterial activity. Carbohydrate Polymers. 2017;**157**:823–833

[24] Christos K, Anastasios E, Ioannis R, Constantinos EE. Lithographically fabricated disposable bismuth-film electrodes for the trace determination of Pb(II) and Cd(II) by anodic stripping voltammetry. Electrochimica Acta. 2008;**53**:5294–5299

[25] Yang D, Wang L, Chen Z, Megharaj M, Naidu R. Voltammetric determination of lead (II) and cadmium (II) using a bismuth film electrode modified with mesoporous silica nanoparticles. Electrochimica Acta. 2014;**132**:223–229

[26] Karim A-Z, Fariba M. Bismuth and Bismuth-Chitosan modified electrodes for determination of two synthetic food colorants by net analyte signal standard addition method. Central European Journal of Chemistry. 2014;**12**:711–718

[27] Leonardi SG, Bonyani M, Ghosh K, Dhara AK, Lombardo L, Donato N, Neri G: Development of a novel Cu(ii) complex modified electrode and a portable electrochemical analyzer for the determination of dissolved oxygen (DO) in water. Chemosensors. 2016;**4**:7–16. DOI: 10.3390/chemosensors4020007

[28] Foster CW, Pillay J, Matters JP, Banks CE. Cobalt phthalocyanine modified electrode utilized in electroanalysis: nono-structured modified electrodes vs. bulk screen-printed electrodes. Sensors. 2014;**14**:21905–21922

[29] Sebarchievici I, Tăranu BO, Birdeanu M, Rus SF, Fagadar-Cosma E. Electrocatalytic behaviour and application of manganese porphyrin/gold nanoparticle- surface modified glassy carbon electrodes. Applied Surface Science. 2016;**390**:131–140

[30] Tung HC, Chooto P, Sawyer DT. Electron-transfer thermodynamics, valence-electron hybridization, and bonding of the meso-tetrakis(2,6-dichlorophenyl)porphinato complexes of manganese, iron, cobalt, nickel, copper, silver, and zinc and of the P+Mn(O) and .bul.P+Fe(O) oxene adducts. Langmuir. 1991;**7**:1635–1641. DOI: 10.1021/la00056a015

[31] Metters JP, Banks CE. Nanoparticle modified electrodes for trace metal ion analysis. In: Honeychurch KC, editor. Nanosensors for Chemical and Biological Applications. 1st ed. Elsevier;2014. p. 54–79. DOI : 10.1533/9780857096722.1.54

[32] Karim-Nezhad G, Khorablou Z, Zamani M, Dorraji PS, Alamgholiloo M. Voltammetric sensor for tartrazine determination in soft drinks using poly (p-aminobenzenesulfonic acid)/zinc oxide nanoparticles in carbon paste electrode. Journal of Food and Drug Analysis. 2016;**xxx**:1–9. (Article in Press, available online 6 November 2016) http://doi.org/10.1016/j.jfda.2016.10.002

[33] Parham H, Zargar B, Shiralipour R. Fast and efficient removal of mercury from water samples using magnetic iron oxide nanoparticles modified with 2-mercaptobenzothiazole. Journal of Hazardous Materials. 2012;**205–206**:94–100

[34] Rannurak J, Sukhotu P, Chooto P. Voltammetric determination of silver(I) using carbon paste electrode modified with 1,8-dihydroxyanthraquinones. In: 205th Meeting of the Electrochemical Society; 9–13 May 2004, San Antonio, Texas, Student Poster Session, Abstract #14.

[35] Serrano N, González-Calabuig A, del Valle M. Crown ether-modified electrodes for the simultaneous stripping voltammetric determination of Cd (II), Pb (II) and Cu (II). Talanta. 2015;**138**:130–137

[36] Gooding JJ, Hibbert DB, Yang W. Electrochemical metal ion sensors. Exploiting amino acids and peptides as recognition elements. Sensors. 2001;**1**:75–90

[37] Serrano N, Prieto-Simón B, Cetó X, del Valle M. Array of peptide-modified electrodes for the simultaneous determination of Pb(II), Cd(II) and Zn(II). Talanta. 2014;**125**:159–166

[38] Ali TA, Mohamed GG, Omar MM, Hanafy NM. Construction and performance characteristics of chemically modified carbon paste electrodes for the selective determination of Co(II) ions in water samples. Journal of Industrial and Engineering Chemistry. 2017;**47**:102–111

[39] Rana S, Mittal SK, Singh N, Singh J, Banks CE. Schiff base modified screen printed electrode for selective determination of aluminium (II) at trace level. Sensors and Actuators B: Chemical. 2017;**239**:17–27

[40] Lopes F, Pacheco JG, Robelo P, Delerue-Matos C. Molecular imprinted elelctrochemical sensor prepared on a screen printed carbon electrode for naloxone detection. Sensor and Actuators B: Chemical. 2017;**243**:745–752

[41] Pourreza N, Rastegarzadeh S, Larki A: Simultaneous preconcentration of Cd(II), Cu(II) and Pb(II) on Nano-TiO_2 modified with 2-mercaptobenzothiazole prior to flame

atomic absorption spectrometric determination. Journal of Industrial and Engineering Chemistry. 2014;**20**:2680–2686

[42] Zhang L, Yang D: Poly(2-mercaptobenzothiazole) modified electrode for the simultaneous determinations of dopamine, uric acid and nitrite. Electrochimica Acta. 2014;**119**:106–113

[43] Fan L, Chen J, Zhu S, Wang M, Xu G. Determination of Cd^{2+} and Pb^{2+} on glassy carbon electrode modified by electrochemical reduction of aromatic diazonium salts. Electrochemistry Communications. 2009;**11**:1823–1825

[44] Innuphat C, Chooto P. Determination of trace levels of Cd(II) in tap water samples by anodic stripping voltammetry with electrografting boron-doped diamond electrode. ScienceAsia. 2017;43:xxx–xxx (Article in press)

[45] Picot M, Lapinsonnière L, Rothballer M, Barrière F. Graphite anode surface modification with controlled reduction of specific aryl diazonium salts for improved microbial fuel cells power output. Biosensors and Bioelectronics. 2011;**28**:181–188

[46] Mooste M, Kibena E, Kozlova J, Marandi M, Matisen L, Niilisk A, Sammelselg V, Tammeveski K. Electrografting and morphological studies of chemical vapour deposition grown graphene sheets modified by electroreduction of aryldiazonium salts. Electrochimica Acta. 2015;**161**:195–204

[47] Bravoa I, García-Mendiola T, Revenga-Parra M, Pariente F, Lorenzo E. Diazonium salt click chemistry based multiwall carbon nanotubeelectrocatalytic platforms. Sensors and Actuators B. 2015;**211**:559–568

[48] Eunkyoung K, Yuan X, Yi C, Hsuan-Chen W, Yi L, et al. Chitosan to connect biology to electronics: Fabricating the bio-device interface and communicating across this interface: A review. Polymers. 2015;**7**:1–46

[49] Carlos AM-H, Carlos CJ, Monica C, Marco Q. Chitosan-modified glassy carbon electrodes: Electrochemical behaviour as a function of the preparation method and pH. Canadian Journal of Analytical Sciences and Spectroscopy. 2009;**54**:53–62

[50] Maghear A, Tertiş M, Fritea L, Marian IO, Indrea E, Walcarius A, Săndulescu R. Tetrabutylammonium-modified clay film electrodes: Characterization and application to the detection of metal ions. Talanta. 2014;**125**:36–44

[51] Wagner J-F, et al. Retention of heavy metals from blast-furnace dedusting sludges by a clayey subsoil. Water, Air, and Soil Pollution. 1991;**1**:351–357. DOI: 10.1007/BF00282898

[52] Liangming W, Nantao H, Yafei Z. Synthesis of polymer—mesoporous silica nanocomposites: A review. Materials. 2010;**3**:4066–4079

[53] Penghui Z, Sheying D, Guangzhe G, Tinglin H. Simultaneous determination of Cd^{2+}, Pb^{2+}, Cu^{2+} and Hg^{2+} at a carbon paste electrode modified with ionic liquid-functionalized ordered mesoporous silica. Bulletin of Korean Chemical Society. 2010;**31**:2949–2954

[54] Abdel Salam OE, Reiad NA, ElShafei MM. A study of the removal characteristics of heavy metals from wastewater by low-cost adsorbents. Journal of Advanced Research. 2011;**2**:297–303

[55] Panumati S, Chudecha K, Vankaew P, Choolert V, Chuenchom L, Innajitara W, Sirichote O. Adsorption efficiencies of calcium (II) and iron (II) ions on activated carbon obtained from pericarp of rubber fruit. Songklanakarin Journal of Science and Technology. 2008;**30**:179–183

[56] Khalil MM, Abed El-aziz GM. Multiwall carbon nanotubes chemically modified carbon paste electrodes for determination of gentamicin sulfate in pharmaceutical preparations and biological fluids. Materials Science and Engineering: C. 2016;**59**:838–846.

[57] Gooding JJ: Nanostructuring electrodes with carbon nanotubes: A review on electrochemistry and applications for sensing. Electrochimica Acta. 2005;**50**:3049–3060

[58] Liang P, Liu Y, Guo L, Zenga J, Lua H. Multiwalled carbon nanotubes as solid-phase extraction adsorbent for the preconcentration of trace metal ions and their determination by inductively coupled plasma atomic emission spectrometry. Journal of Analytical Atomic Spectrometry. 2004;**19**:1489–1492. DOI: 10.1039/B409619C

[59] Tuzena M, Saygia KO, Soylakb M. Solid phase extraction of heavy metal ions in environmental samples on multiwalled carbon nanotubes. Journal of Hazardous Materials. 2008;**152**:632–639

[60] Rao GP, Lu C, Su F. Sorption of divalent metal ions from aqueous solution by carbon nanotubes: A review. Separation and Purification Technology. 2007;**58**:224–231

[61] Che G, Lakshmi BB, Fisher ER, Martin CR. Carbon nanotubule membranes for electrochemical energy storage and production. Nature. 1998;**393**:346–349. DOI: 10.1038/30694

[62] Pérez-Ràfols C, Serrano N, Díaz-Cruz JM, Ariño C, Esteban M. Glutathione modified screen-printed carbon nanofiber electrode for the voltammetric determination of metal ions in natural samples. Talanta. 2016;**155**:8–13

[63] María-Hormigos R, Gismera MJ, Procopio JR, Teresa Sevilla MT. Disposable screen-printed electrode modified with bismuth–PSS composites as high sensitive sensor for cadmium and lead determination. Journal of Electroanalytical Chemistry. 2016;**767**:114–122.

[64] Haeidari H, Habibi B, Vaigan FB. Glassy carbon electrode modified with an ordered mesoporous carbon/Ag nanoparticle nanocomposite for the selective detection of iodate. Analytical Methods. 2016;**8**:4406–4412. DOI: 10.1039/c6ay01087c

[65] March G, Nguyen TD, Piro B. Modified electrodes used for electrochemical detection of metal ions in environmental analysis: A review. Biosensors. 2015;**5**:241–275

[66] Johri N, Jacquillet G, Unwin R. Heavy metal poisoning: the effects of cadmium on the kidney. Biometals. 2010;**23**:783–792

[67] http://water.epa.gov

[68] Chooto P, Wararattananurak P, Kangkamano T, Innuphat C, Sirinawin W. Determination of inorganic arsenic species by hydride generation atomic absorption spectrophotometry and cathodic stripping voltammetry. ScienceAsia. 2015;**41**:187–197. DOI: 10.2306/scienceasia1513-1874.2015.41.187

[69] Chuaysong R, Chooto P, Pakawatchai C. Electrochemical properties of copper(I) halides and substituted thiourea complexes. ScienceAsia. 2008;**34**:440–442. DOI: 10.2306/scienceasia1513–1874.2008.34.440

[70] Tapachai WA, Vataporna S, Pakawatchai C, Chooto P, Innuphat C. Synthesis and characterization of bis(2-mercaptobenzimidazole)bromo- and iodocopper(I) complex. ScienceAsia. 2017;43:xxx–xxx (Article in press)

[71] Thanapackium P, Rameshkumar S, Subramanian SS, Mallaiya K. Electrchemical evaluation of inhibition efficiency of ciprofloxacin on the corrosion of copper in acid media. Materials Chemistry and Physics. 2016;**174**:129–137

[72] Yu C, Gou L, Zhou X, Bao N, Gu H. Chitosan-Fe_3O_4 nanocomposite based electrochemical sensors for the determination of bisphenol A. Electrochimica Acta. 2011;**56**:9056–9063

[73] Li Y, Zhai X, Liu X, Wang L, Liu H, Wang H. Electrochemical determination of bisphenol A at ordered mesoporous carbon modified nano-carbon ionic liquid paste electrode. Talanta. 2016;**148**:362–369

[74] Hou C, Tang W, Zhang C, Wang Y, Zhu N. A novel and sensitive electrochemical sensor for bisphenol A determination based on carbon black supporting ferroferric oxide nanoparticles. Electrochimica Acta. 2014;**144**:324–331

[75] Niu X, Yang W, Wang G, Ren J, Guo H, Gao J. A novel electrochemical sensor of bisphenol A based on stacked graphene nanofibers/gold nanoparticlescomposite modified glassy carbon electrode. Electrochimica Acta. 2013;**98**:167–175

[76] Zhou L, Wang J, Li D, Li Y. An electrochemical aptasensor based on gold nanoparticles dotted graphene modified glassy carbon electrode for label-freedetection of bisphenol A in milk samples. Food Chemistry. 2014;**162**:34–40

[77] Yin H, Cui L, Chen Q, Shi W, Ai S, Zhu L, Lu L. Amperometric determination of bisphenol A in milk using PAMAM–Fe_3O_4 modified glassy carbon electrode. Food Chemistry. 2011;**125**:1097–1103

[78] Wang Y, Cokeliler D, Gunasekaran S: Reduced graphene oxide/carbon nanotube/gold nanoparticles nanocomposite functionalized screen-printed electrode for sensitive electrochemical detection of endocrine disruptor bisphenol A. Electroanalysis. 2015;**27**:2527–2536

[79] Cerovac S, Guzsvány V, Kónya Z. Trace level voltammetric determination of lead and cadmium in sediment pore water by a bismuth-oxychloride particlemultiwalled carbon nanotube composite modified glassy carbon electrode. Talanta. 2015;**134**:640–649

[80] Zhou W, Li C, Sun C, Yang X. Simultaneously determination of trace Cd^{2+} and Pb^{2+} based on L-cysteine/graphene modified glassy carbon electrode. Food Chemistry. 2016;**192**:351–357

[81] Chamjangali MA, Kouhestani H, Masdarolomoor F, Daneshinejad H. A voltammetric sensor based on the glassy carbon electrode modified with multi-walled carbon nanotube/poly(pyrocatechol violet)/bismuth film for determination of cadmium and lead as environmental pollutants. Sensors and Actuators B: Chemical. 2015;**216**:384–393

[82] Wu Y, Li NB, Luo HQ. Simultaneous measurement of Pb, Cd and Zn using differential pulse anodic stripping voltammetry at a bismuth/poly (paminobenzene sulfonic acid) film electrode. Sensors and Actuators B: Chemical. 2008;**133**:677–681

[83] Dimovasilis PA, Prodromidis MI. Bismuth-dispersed xerogel-based composite films for trace Pb (II) and Cd (II) voltammetric determination. Analytica Chimica Acta. 2013;**769**:49–55. DOI: 10.1016/j.aca.2013.01.040

[84] Injang U, Noyrod P, Siangproh W. Determination of trace heavy metals in herbs by sequential injection analysis-anodic stripping voltammetry using screen-printed carbon nanotubes electrodes. Analytica Chimica Acta. 2010;**668**:54–60. DOI: 10.1016/j.aca.2010.01.018

[85] Marinho JZ, Silva RAB, Barbosa TGG. Graphite-composite electrodes bulk-modified with $(BiO)_2CO_3$ and Bi_2O_3 plates-like nanostructures for trace metal determination by anodic stripping voltammetry. Electroanalysis. 2013;**25**:765–770. DOI: 10.1002/elan.201200592

[86] Sopha H, Baldrianová L, Tesarová E. A new type of bismuth electrode for electrochemical stripping analysis based on the ammonium tetrafluorobismuthate bulk-modified carbon paste. Electroanalysis. 2010;**22**:1489–1493. DOI: 10.1002/elan.201070010

[87] Ding Y, Wang Y, Li BK, Lei Y. Electrospun hemoglobin microbelts based biosensor for sensitive detection of hydrogen peroxide and nitrite. Biosensors and Bioelectronics. 2010;**25**:2009–2015

[88] Ma LP, Yuan R, Chai YQ, Chen SH. Amperometric hydrogen peroxide biosensor based on the immobilization of HRP on DNA-silver nanohybrids and PDDA-protected gold nanoparticles. Journal of Molecular Catalysis B: Enzymatic. 2009;**56**:215–220

[89] Salimi A, Noorbakhsh A, Mamkhezri H, Ghavami R. Electrocatalytic reduction of H_2O_2 and oxygen on the surface of thionin incorporated onto MWCNTs modified glassy carbon electrode: Application to glucose detection. Electroanalysis. 2007;**19**:1100–1108. DOI: 10.1002/elan.200603828

[90] Feng JJ, Zhao G, Xu JJ, Chen HY. Direct electrochemistry and electrocatalysis of heme proteins immobilized on gold nanoparticles stabilized by chitosan. Analytical Biochemistry. 2005;**342**:280–286

[91] Wang S, Xie F, Liu G: Direct electrochemistry and electrocatalysis of heme proteins on SWCNTs-CTAB modified electrodes. Talanta. 2009;**77**:1343–1350

[92] Zhu JT, Shi CG, Xu JJ, Chen HY. Direct electrochemistry and electrocatalysis of hemoglobin on undoped nanocrystalline diamond modified glassy carbon electrode. Bioelectrochemistry. 2007;**71**:243–248

[93] Shan D, Cheng G, Zhu D, Xue H, Cosnier S, Ding S. Direct electrochemistry of hemoglobin in poly(acrylonitrile-co-acrylic acid) and its catalysis to H_2O_2. Sensors and Actuators B: Chemical. 2009;**137**:259–265.

[94] Shan D, Wang S, Xue H, Cosnier S. Direct electrochemistry and electrocatalysis of hemoglobin entrapped in composite matrix based on chitosan and $CaCO_3$ nanoparticles. Electrochemistry Communications. 2007;**9**:529–534

[95] Zhang JD, Oyama M. A hydrogen peroxide sensor based on the peroxidase activity of hemoglobin immobilized on gold nanoparticles-modified ITO electrode. Electrochimica Acta. 2004;**50**:85–90

[96] Xu YX, Hu CG, Hu SS. A hydrogen peroxide biosensor based on direct electrochemistry of hemoglobin in Hb-Ag sol films. Sensors and Actuators B: Chemical. 2008;**130**:816–822

[97] Yu CM, Zhou XH, Gu HY. Immobilization, direct electrochemistry and electrocatalysis of hemoglobin on colloidal silver nanoparticles-chitosan film. Electrochimica Acta. 2010;**55**:8738–8743

[98] Aldrich AP, van den Berg CM. Determination of iron and its redox speciation in seawater using catalytic cathodic stripping voltammetry. Electroanalysis. 1998;**10**:369–373

[99] Laglera LM, Santos-Echeandia J, Caprara S, Monticelli D. Quantification of iron in seawater at the low picomolar range based on optimization of bromate/ammonia/dihydroxynaphtalene system by catalytic adsorptive cathodic stripping voltammetry. Analytical Chemistry. 2013;**85**:2486–2492

[100] Gun J, Salaun P, van den Berg CM. Advantages of using a mercury coated micro-wire, electrode in adsorptive cathodic stripping voltammetry. Analytica Chimica Acta. 2006;**571**:86–92

[101] Zhao J, Jin W. A study on the adsorption voltammetry of the iron(III)-2-(5′-bromo-2′-pyridylazo)-5-diethylaminophenol system. Electroanalytical Chemistry. 1989;**267**: 271–278

[102] Li F, Pan D, Lin M, Han H, Hu X, Kang Q. Electrochemical determination of iron in coastal waters based on ionic liquid-reduced graphene oxide supported gold nanodendrites. Electrochimica Acta. 2015;**176**:548–554

[103] Inam R, Somer G. A direct method for the determination of selenium and lead in cow's milk by differential pulse stripping voltammetry. Food Chemistry. 2000;**69**:345–350

[104] Zhang Q, Li X, Shi H, Hongzhou, Yuan Z.Determination of trace selenium by differential pulse adsorptive stripping voltammetry at a bismuth film electrode. Electrochimica Acta. 2010;**55**:4717–4721

[105] Ramadan AA, Mandil H, Shikh-Debes A. Differential pulse anodic stripping voltammetric determination of selenium(IV) at a gold electrode modified with 3, 3′-diaminobenzidine·4HCl-nafion. International Journal of Pharmacy & Pharmaceutical Sciences. 2014;**6**:148–153

[106] Kolliopoulos AV, Metters JP, Banks CE. Electroanalytical sensing of selenium(IV) utilising screen printed graphite macro electrodes. Analytical Methods. 2013;**5**:851–856

[107] Fierro S, Watanabe T, Akai K, Yamanuki M, Einaga Y. Anodic stripping voltammetry of Se^{4+} on gold-modified boron-doped diamond electrodes. International Journal of Electrochemistry. 2012;**2012**:1–5

[108] Cai QT, Khoo SB. Poly (3 3′-diaminobenzidine) film on a gold electrode for selective preconcentration and stripping analysis of selenium (IV). Analytical Chemistry. 1994;**66**:4543–4550

[109] Baś B, Jedlińska K, Węgiel K. New electrochemical sensor with the renewable silver annular band working electrode: fabrication and application for determination of selenium(IV) by cathodic stripping voltammetry. Electrochemistry Communications. 2014;**49**:79–82

[110] Segura R, Pizarro J, Díaz K, Placencio A, Godoy F, Pino E, Recioc F. Development of electrochemical sensors for the determination of selenium using gold nanoparticles modified electrodes. Sensors and Actuators B: Chemical. 2015;**220**:263–269

[111] Gao D, Cai F, Wang G, Bao X. Nanostructured heterogeneous catalysts for electrochemical reduction of CO_2. Current Opinion in Green and Sustainable Chemistry. 2017;**3**:39–44

[112] Horn EJ, Rosen BR, Baran PS. Synthetic organic electrochemistry: an enabling and innately sustainable method. ACS Central Science. 2016;**2**:302–308. DOI: 10.1021/acscentsci.6b00091

[113] Hernández CN, García MBG, Santos DH, Heras MA, Colina A, Fanjul-Bolado P. Aqueous UV-VIS spectroelectrochemical study of the volatammetric reduction of grapheme oxide on screen-printed carbon electrodes. Electrochemistry Communications. 2016;**64**:65–68